# Total Productive Maintenance
# The Western way

# Total Productive Maintenance
# The Western way

*Peter Willmott*

Butterworth-Heinemann
Linacre House, Jordan Hill, Oxford OX2 8DP
225 Wildwood Avenue, Woburn, MA 01801–2041
A division of Reed Educational and Professional Publishing Ltd

℞ A member of the Reed Elsevier plc group

OXFORD  AUCKLAND  BOSTON
JOHANNESBURG  MELBOURNE  NEW DELHI

First published 1994
Reprinted 1997, 1999

**British Library Cataloguing in Publication Data**
Willmott, Peter
  Total Productive Maintenance: Western Way
  I. Title
  658.202

ISBN 0 7506 1925 2

**Library of Congress Cataloguing in Publication Data**
Willmott, Peter
  Total productive maintenance. The Western way/Peter Willmott.
  p.    cm.
  Includes bibliographical references and index.
  ISBN 0 7506 1925 2
  1. Plant maintenance – Management. 2. total productive maintenance.
  I. title
  TS 192.W53                                         94–12560
  658.2'02–dc20                                          CIP

Typeset and illustrated by TecSet Ltd, Wallington, Surrey
Printed and bound in Great Britain by Athenæum Press Ltd,
Gateshead, Tyne & Wear

# Contents

# Foreword

Three years ago I joined the Royal Mail as Engineering Director. As I travelled around the organization I realized that we urgently needed a new strategy for Engineering, and in particular, maintenance. In common with many other major organizations whose primary goal is not necessarily related to technology, maintenance was an issue which did not engage most people with enthusiasm, excitement or interest. It was some months before I discovered TPM at a full-day seminar staged by the Institution of Mechanical Engineers.

This was a new way of placing maintenance, and many other related issues, in a proper context within a total quality organization. As I investigated further it was only a short time before I met Peter Willmott. At that time Peter was one of a tiny number of consultants, and experts, in the UK who knew something specific about TPM. Peter was able to introduce me to a handful of companies with whom he had worked recently, and together they finally convinced me that TPM was a suitable major strategy for re-inventing Engineering within the Royal Mail.

Since then things have changed dramatically. TPM has been adopted by the Royal Mail as the major way of caring for its considerable technological assets. The implementation will probably be the largest in the country, and will certainly be at the core of our advanced automation strategy.

A year after starting the TPM journey in Royal Mail I joined Peter Willmott and a small number of enthusiasts to found the UK TPM club at the Institution of Mechanical Engineers. This has grown rapidly and now has a membership of over 50 companies. It supports a wide range of activities aimed at introducing more organizations and individuals, to the benefits of TPM. Peter has remained one of the topic's major promoters and has increasingly become recognized as one of the real experts in this specific area. His own business has grown from a small personal consultancy to an organization supporting dozens of companies in studying the potential of TPM, preparing for launch, and full implementation programmes. His initial efforts in the UK have now spread across most European countries and many companies now owe him a considerable debt in terms of the improvement in quality, effectiveness and performance he has helped them to achieve.

Peter's book is a milestone in TPM in the UK because it will enable far more people to become aware of the approach. Although its popularity has increased considerably in the last three years TPM is still a minority subject compared to its more mature counterparts such as total quality. However, I am convinced that the organizations who could not make use of the TPM approach either directly, or in a modified form are few and far between. This book helps to take the original material, which was published in Japanese and then translated into English in North America, and represents it in a style more suitable for western European companies in the mid 1990s. The book avoids the dangers of becoming academic, while still including considerable intellectual content.

Please consider the material in the book carefully. Our own journey into TPM has made us realize that many of our basic assumptions about small groups in the workplace, defect avoidance, kaizen and other issues could be improved with the new insights we have gained. The book includes some of the best practical stories, presentational techniques, and aids to understanding which Peter has developed over the past years, and they will be just as valuable to you in explaining TPM.

The most important thing in the book is the visible evidence of the enthusiasm Peter and I share for TPM, and the excitement we want you to feel in discovering such an effective tool for performance and quality improvement.

Dr Duncan Hine
Engineering Director, Royal Mail
Chairman, UK TPM Club
July 1994

# Preface

Customers expect manufacturers to provide excellent quality, reliable delivery and competitive pricing. This demands that the manufacturer's machines and processes are highly reliable. But what does this phrase 'highly reliable' really mean?

Certainly, with manufacturing, process and service industries becoming progressively dependent on the reliability of fewer but more sophisticated machines and processes, poor equipment operating performance is no longer affordable or acceptable. The overall effectiveness of our machines, equipment and processes is paramount to provide consistency of product quality and supply at a realistic price.

Coping with the modern manufacturing technology that is intrinsic in the materials, mechanisms and processes which we invent, design and use is one issue. Delivering that manufacturing company's vision and values as a lean, just-in-time producer to its customers, shareholders and employees is another.

Some world class Japanese companies recognized over twenty years ago that the effective application of modern technology can only be achieved through people – starting with the operators of that technology – and not through systems alone. Hence the emergence of total, productive, maintenance or TPM. Total productive maintenance is the enabling tool to maximize the effectiveness of our equipment by setting and maintaining the optimum relationship between people and machines.

The problem with the words 'total productive maintenance' – and hence the philosophy or technique of TPM – is that, to Western ears, they sound as though TPM is a maintenance function or a maintenance department initiative. But it is *not*! On the contrary, TPM is driven by manufacturing which picks up production and maintenance as equal partners: it is no longer appropriate to say 'I operate, you fix' and 'I add value, you cost money.' What TPM promotes is: 'We are *both* responsible for this machine, process or equipment and, between us, *we* will determine the best way to operate, maintain and support it.'

The problem of definition has arisen because the word 'maintenance' has a much more comprehensive meaning in Japan than in the Western world. If you ask someone from a typical Western manufacturing company to define the word 'maintenance', at best he might say, 'Carry out planned servicing at fixed intervals'; at worst he might say, 'Fix it when it breaks down.' If you ask a Japanese person from a world class manufacturing company he will probably say, 'Maintenance means maintaining and *improving* the integrity of our production and quality systems through the machines, processes, equipment and people who add value to our products and services, that is the operators *and* maintainers of our equipment.' Whilst this may be a longer definition it is also a more comprehensive and relevant description.

Over the last few years, certainly since the advent of the 1990s, a growing number of Western companies have, with varying degrees of success, adopted the Japanese TPM philosophy. The companies who have been successful in using TPM in their operations have recognized and applied some key success factors including:

- TPM is led by manufacturing.
- TPM is a practical application of total quality and teamwork.
- TPM is an empowerment process.
- The TPM philosophy is like a heart transplant: if you don't match it to the patient you will get rejection. You must therefore treat each company or recipient as unique and *adapt* the principles of TPM to suit the local plant-specific issues.

This book provides a comprehensive introduction to TPM by taking the reader through the evolution and development of TPM and then basing the reality on the actual 'making it happen' element of TPM, taking account of differing Western cultures and industries. It is intended for manufacturing, operations and maintenance management practitioners who want to understand TPM and what it can achieve and to know how to set about applying it. It will also provide students of engineering, management and business studies with an insight into this potentially powerful philosophy which make appropriate use of a number of proven tools and techniques.

Chapters 1 to 4 explain the background to TPM and its Japanese origins. They give an overview of how TPM is an enabling tool to improve the overall effectiveness of equipment and processes, as well as a demonstration of total quality, teamwork and empowerment in action. Chapters 5 and 6 provide the detail of the TPM process as developed from the original Japanese approach without destroying Nakajima's five principles, but rather reflecting our differing Western perceptions and culture.

Chapters 7 and 8 provide a management perspective on how to plan and launch a TPM programme and then to measure and sustain the commitment beyond the introductory pilot stages to the ultimate goal of flawless operation. Chapter 9 outlines the perspective of TPM for designers and equipment planners, on the basis that it makes sense to strive for the notion of flawless operation in the design and specification phase rather than five years after the equipment has been on the shop floor! Finally, Chapters 10–14 provide some live case studies taken from a variety of industrial sectors.

PETER WILLMOTT

# Acknowledgements

When I was first approached about writing this book, some years ago, it seemed a splendid idea. However, as I planned the book, the enormity of the task quickly dawned on me. I could talk about TPM for hours and hours: in fact my partner in WCS International, Dennis McCarthy, has often said, 'Will the last person leaving the room please switch off Peter Willmott!' Writing is a vastly different matter, and I have never pretended to be a good wordsmith. On the contrary, I spend endless and valuable time worrying about a sentence I have written because 'it's not quite right'; the problem is that I cannot think of the right bit!

I am therefore particularly indebted to Peter Osborn, who has taken everything I have thrown at him – dictaphone tapes, video tapes, articles, visual aids and scribblings – and come up with a logical structure and, I hope, an interesting book for you, the reader. Peter Osborn has four priceless assets:

- He understands engineering as a qualified engineer.
- He is technical editor of the *Plant and Works Engineering* journal.
- He has studied and written about Japanese techniques and philosophies of just in time, continuous improvement and total quality.
- Finally, although in his early seventies, he has boundless energy!

Any decent technical book only has lasting substance if it is based on practice rather than just theory. Here I would like to express my thanks to the many hundreds of people I have come into contact with during the course of my work, particularly over the last five years or so.

When I was with March Consulting Group, I headed a significant study for the UK's Department of Trade and Industry aimed at identifying maintenance best practice in industry. Peter Connett, now retired from the DTI, was a particularly close ally in promoting the importance of maintenance to industry, and I still make regular use of his 'healthy body, healthy equipment' analogy in explaining the principles of TPM.

David Miles and Kris Larsen of the European Union BRITE-EURAM programme in Brussels also provided demanding but stimulating encouragement when I headed a similar study to look into maintenance techniques and technologies throughout the twelve member states of the EU.

Our many clients, who have proved and are continuing to prove the value of correctly tailored TPM programmes in their own companies, are really the cornerstone of live TPM experiences. I am particularly indebted to:

- Warren Burgess, John Wiseman and Les Thompson of BP Exploration
- Roger Olney and John Wright of Castle Cement
- Tony Andrews, Paul Grozier and Kevin Smither of Abbott Laboratories

- Tim Holton, David Espley, Jeremy Paul and Graham James of Courtaulds, Swindon
- John Hurst, Mike Chapman, Albert Lidauer and Manfred Ullrich of General Motors Europe
- Ian Ross and Dave Dixon of Vauxhall Motors
- Lynn Williams of the EETPU.

In spreading the TPM message via conferences, workshops and study tours, the roles played by David Willson, John Moulton, Tom Brock, Peter Pugh and Lisa Hearst are all acknowledged with gratitude.

Professor Yamashima of Kyoto University, a PM Excellence Award Examiner for the Japan Institute of Plant Maintenance, has been a constant source of inspiration for me. Similarly, the subject of TPM would not have been given its clarity, structure and boundaries had it not been for Seiici Nakajima who pioneered the approach in Japan over twenty years ago. As such I am eternally grateful for his early sources of TPM knowledge and understanding.

A special word of thanks is due to my colleagues in WCS International who have provided much of the inspiration, perspiration and material for this book.

Thanks also to Duncan Hine for his kind contribution in writing the Foreword to this book.

Last, but not least, I am grateful to my wife for having the patience to support me and for creating an environment which has given me the time to write this book.

# 1

# Background and Evolution of TPM

The father of total productive maintenance (TPM) was Seiici Nakajima, who pioneered the approach in Japan and exerted a major influence over the economic progress made by Japanese manufacturers from the late 1970s.

The achievement of world class manufacturing performance is essential if companies are to survive against international competition. One vital ingredient of this is highly reliable and consistently operated machines, equipment and processes. It is perhaps worth reminding ourselves that these machines, equipment and processes together with their operators and maintainers are the only direct wealth creators in a manufacturing plant; all other functions such as sales, marketing, design, production control and finance either feed off them or support them. To build a consistent and sustainable competitive advantage it is essential that the man/machine interface is maximized and that all associated sources of waste are eliminated.

Total productive maintenance is based on teamwork and provides a method for the achievement of world class levels of *overall equipment effectiveness* through people and not through technology or systems alone. A growing number of Western companies are enthusiastically adopting the common-sense approach of TPM, recognizing that it is both a world class process and a grass roots process with measurable benefits. It is total quality with teeth! The phrase 'total productive maintenance' has been widely adopted to describe the process, but in Western terms it is a misnomer, implying that it is maintenance driven: it certainly is not! In the West it is a manufacturing-led initiative driven by production and maintenance as equal partners. The definition of maintenance in Japan is 'Maintaining and *improving* the integrity of our production systems through the machines, equipment, processes and employees that add value'; this contrasts vividly with the traditional role of the maintenance department in the West, i.e. 'Fix it when it breaks down'! Figure 1.1 compares the approaches to maintenance of a world class Japanese company and a traditional Western company.

Nakajima's teachings rest on five pillars:

1   Adopt improvement activities designed to increase the overall equipment effectiveness (OEE) by attacking the six losses.
2   Improve existing planned and predictive maintenance systems.
3   Establish a level of self-maintenance and cleaning carried out by highly trained operators.
4   Increase the skills and motivation of operators and engineers by individual and group development.
5   Initiate maintenance prevention techniques including improved design and procurement.

| Japanese | Western |
|---|---|
| • Manufacturing led | • Functional/departmental |
| • Versatile, flexible, multi-skilled shop floor teams | • Demarcations |
| • Predictive, preventive problem elimination | • Reactive maintenance |
| • Lost opportunity benefits | • Direct costs |
| • Vitally important | • Unnecessary evil |
| • Shop floor ownership | • Production management dictate |

**Figure 1.1**   *Comparison of Japanese and Western approaches to maintenance*

The five principles which are central to the approach of Western TPM have the same objectives but are treated by WCS International in a manner more appropriate to Western methods and culture. Hence one of the main purposes of this book is to adapt the Japanese approach without losing sight of Nakajima's principles. It is rather like a heart transplant operation: if you do not match the donor heart to the recipient you will get rejection. The relationship between the Japanese model of TPM and the Western way is fully covered in Chapter 4.

The concept of *total productive maintenance* is simple. TPM seeks to engender a company-wide approach towards achieving a standard of performance in manufacturing, in terms of the overall effectiveness of equipment, machines and processes, which is truly world class. It does not, however, rest at that point, but strives for continuous improvement aimed at achieving and sustaining flawless operation.

The philosophy at the heart of the TPM process is that all the assets on which production depends are kept always in optimum condition and available for maximum output. This is made possible because those using the plant and equipment are personally and directly involved in ensuring that this happens.

There are three particular features in the TPM process:

- TPM embraces methods for data collection, analysis, problem solving and process control – methods which aim always at improving equipment effectiveness.
- Being led by manufacturing, TPM encourages production (process) and maintenance (engineers) to work together as equal partners. It also involves other departments such as design, quality, production control, finance and purchasing who are concerned with equipment; this of course also includes management and supervision.

- TPM fosters the continuous improvement of equipment and in doing so makes extensive use of standardization, workplace organization, visual management and problem solving.

The cost of a maintenance department can readily be assessed from hours, overheads, materials and subcontract work, and standard accounting practices will produce these costs. However, the cost/benefit equation of maintenance work is much more difficult to assess. Loss of output due to a major breakdown can be quantified, but the effects of minor stoppages, reduced running speed, idling, quality defects, yield deficiencies and startup losses are hard to measure and consequently are rarely valued. TPM provides an asset care regime which aims to improve overall equipment effectiveness by resolving equipment-related problems *once and for all*. As TPM proceeds, the unmeasured shortcomings are progressively reduced and a true cost/benefit equation can be evaluated.

Looking back to the era after the First World War, the origins of manufacturing efficiency began with the mass production philosophy of the great American pioneers such as Henry Ford. This led in due course to the 'economic batch' which justified the tooling and setup and which brought with it first rate fixing and later the trappings of work study and performance-related pay.

In the rapidly changing era after the Second World War, products became more complex and the propensity to produce scrap gained increasing significance. Parallel with this, production control became increasingly difficult and the early manual systems became accident prone and created endless frustration on the production lines. All this led to broken despatch time promises and disgruntled customers. During this era the Japanese were rebuilding their economy from scratch and, through their post-war associations with the USA, they took the best ideas on efficient production, improved on them and adapted them for their own use. The culmination of this process was seen in the achievements of the great Japanese industrialists such as Taiichi Ohno of Toyota. From the middle 1970s Japanese consultants travelled widely in the Western world and propagated their own approaches to manufacturing efficiency. The real spur to Japanese industrial efficiency came from the oil shock in the late 1970s when they found themselves faced with rapidly rising energy costs and no indigenous sources of energy at all. The concept of total productive maintenance was pioneered by Seiici Nakajima.

This book seeks to present TPM in a manner which takes account of the fundamental differences in culture between Japan and the Western world. It further recognizes that generalizations do not apply even to the Western world: an approach which will succeed in Germany may not be right for France, Spain or Belgium. There are likewise culture differences between such organizations as ICI, General Motors, Courtaulds and the Royal Mail. TPM will succeed only if it is tailored to the unique, plant-level circumstances where it has to develop and thrive.

Some of the main differences between the work culture in *world class* Japanese companies and that which has traditionally prevailed in traditional Western companies are as follows:

- The work ethic is number one in Japan and comes before self and family.
- The Japanese *belong* to enterprises rather than work for companies.
- They are accustomed to the idea of reaching a consensus and to the principle of empowerment of the individual rather than 'them and us' or dictate from the bosses.
- They plan for action rather than react to events.
- They seek stages, phases, pillars and steps rather than just 'go and do it'.
- They strive for long-term sustainable improvements, not instant results.
- Activities are benefit driven, not cost constrained.
- The Japanese believe that the necessary input will result in the desired output and are not preoccupied with the downside.
- They are inquisitive and keen to try new ideas rather than to preserve the status quo.
- They pay full attention to detail rather than use a broad brush approach.

The relationship between just in time (JIT) (also of Japanese origin) and TPM needs to be clearly understood. JIT rests on five major factors:

- Encouraging the ability, energy, intelligence and enthusiasm of the workforce by inspiring them to increase productivity through working more effectively and hence more efficiently.
- Reducing the level of stock and work in progress with rigour and without relent to the point where nothing is manufactured unless it fits into a production programme designed to meet the specific needs of a customer. There is no more and no less manufacture than that required to meet those needs.
- Abandoning the concept of mass production or even batch production in favour of the approach known as 'one-piece flow'. This aims to produce only what is needed, in the shape, style and finish needed, hour by hour and minute by minute. To achieve this, reliance must be placed on the versatility and reliability of production machinery and on the speed and ease with which setup and changeover can be achieved using judiciously applied technology and guided by well-trained, motivated and therefore experienced people.
- Shortening production lines to an extent which is hardly believable, and adopting U-shaped lines so that operators can carry out many different processes according to the needs of the moment.
- Taking pride in eliminating all sources of waste so that the business prospers in the interests of all. Operators are positioned so that everything they need is to hand, so that they are mobile and so that nothing stands in the way of furthering the production process with the constant objective of satisfying the customer.

Just in time will not work *unless* you have highly reliable and effective equipment, where the interface between people and machines is maximized – which is a major objective of TPM.

JIT uses many Japanese expressions which have become the essential language of JIT, and many of these expressions are also an essential part of TPM. The language of Western TPM is developed in Chapter 3, and this language is a vital part of training for organizations adopting TPM.

Over the last 30 years there have been many different approaches to improving maintenance efficiency, and these are briefly summarized as follows.

## Breakdown maintenance

Breakdown maintenance rested on the concept of a highly skilled and dedicated maintenance team poised ready to step in when plant failed for whatever reason. The maintenance team took pride in their ability to 'fix' a broken-down machine and in their forethought in squirrelling away the parts and the special tools needed for the job. The GITAFI regime prevailed – get in there and fix it!

## Planned preventive maintenance

After the Second World War the increase in the number, variety and complexity of physical assets – plant, equipment and buildings – necessitated a different approach. At the same time the relationship of plant failure to safety, environment, product quality and, above all, cost dictated the vital necessity to achieve much higher plant availability. Planned preventive maintenance (PPM) rested on overhauls and checks at fixed intervals which were applied on a scheduled maintenance basis.

## Predictive maintenance

Predictive maintenance (PM) followed, and this rested heavily on systems of asset and failure records supported in the early days by cumbersome computers and somewhat primitive software. These limitations meant that the results of the data collection were not used and systems frequently fell into disrepute because of lack of data integrity and the absence of regular and disciplined data collection.

## Computer-aided maintenance

Computer-aided maintenance (CAM) developed in parallel with PM, and with it came smaller portable computers with improved user-friendly software packages.

## Quality assurance

During this period, inspection of parts and finished products gave way to quality control and quality assurance as more sensible, predictive approaches to achieving product quality.

## Condition-based monitoring

With the 1980s came a similar emphasis within the maintenance function to predict equipment (as opposed to product) condition. Condition-based monitoring (CBM) is now a highly developed practical aid to effective maintenance. It rests on three basic approaches:

- vibration analysis aided by sophisticated software, enabling the signs of failures to be detected at a very early stage
- analysis of oils and hydraulic fluids to detect signs of wear or threats to the integrity of hydraulic systems
- thermography, primarily for the early detection of electrical and electronic faults.

Also in the 1980s there was a major shift towards multi-skilled tradesmen and some elements of teamwork within the maintenance function.

## Reliability-centred maintenance

Reliability-centred maintenance (RCM) had its origins in the aircraft industry, and is also especially relevant where public health and safety may be at risk, notably in nuclear installations, oil tankers, offshore oil platforms and chemical manufacturing plants. RCM makes extensive use of statistical and mathematical techniques to predict reliability and to assess maintainability. By its nature RCM is maintenance driven because failure *can* be catastrophic. A good illustration is the exterior of an aircraft in flight: the pilot is alerted to a malfunction only through his instruments, for he cannot walk out on a wing and see for himself!

## Total Quality Management

Total quality management (TQM) rests on the elevation of the inspection function within a company to a commitment to achieve and adhere to one of the three main models of the ISO 9000 standard, *and* to treat *everybody* (internal and external) as customers. The TPM process embodies, as an essential, the achievement of total quality and is the logical outcome of these developments in the maintenance function over the years.

TPM uses the word 'empowerment' to signify the power of the operator and the maintainer, as members of a team, to ensure that the assets they 'own' in the shape of equipment and machinery are kept always at maximum effectiveness and under permanent scrutiny for continuous improvement. The concept of continuous improvement is embodied in the Japanese word *kaizen*, which is one of the keys to Japanese success and which has been developed and presented to the Western world by Masaaki Imai.

All this leads to the typical *core* business objectives of TPM:

- customer satisfaction
- world class performance
- cost competitiveness
- market share.

All are achieved through highly reliable and effective equipment.

There is no better way of summarizing the content of this chapter than to refer to Figures 1.1 to 1.5. Figure 1.2 illustrates that in order to achieve a total quality production system of world class stature you have materials with value being added through methods and machines and the common denominator of people. TPM provides the link between people and their machines. Figure 1.3 shows how TPM has evolved from earlier approaches. Figure 1.4 is a further reminder of the need to motivate people who are at the centre of the business driver process. Finally, Figure 1.5 shows how the TPM process leads to world class performance.

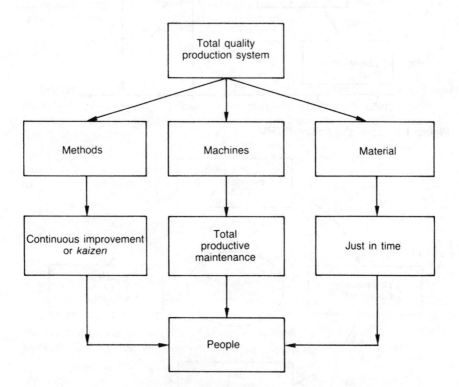

**Figure 1.2** *Structure of a total quality production system*

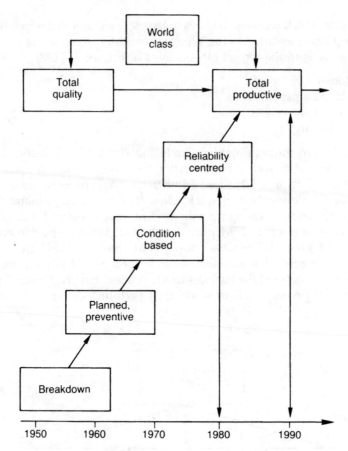

**Figure 1.3** *Emerging power of TPM*

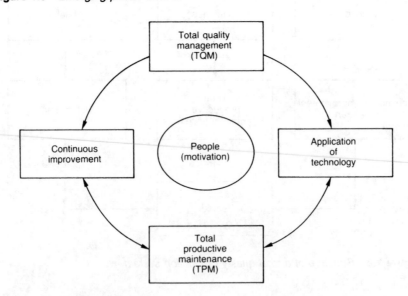

**Figure 1.4** *Business drivers*

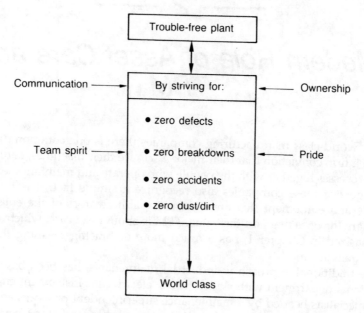

**Figure 1.5** *Factors in world class performance*

# 2

# Modern Role of Asset Care and TPM

In the world class manufacturing companies there is one common denominator: a firm conviction that their major *assets* are their machines, equipment and processes together with the people who operate and maintain them. The managers of these companies also recognize a simple fact: it is the same people and equipment that are the true wealth creators of the enterprise. They are the ones that add the value. TPM is about asset care, which, as was emphasized in Chapter 1, has a much more embracing meaning than the word 'maintenance'.

The traditional approach to industrial maintenance has been based on a functional department with skilled fitters, electricians, instrument engineers and specialists headed by a maintenance superintendent or works engineer. The department was supported by its own workshop and stores containing spares known from experience to be required to keep the plant running. The maintenance team would take great pride in its ability to 'fix' a breakdown or failure in minimum time, working overnight or at weekends and achieving the seemingly impossible. Specialized spares and replacements would be held in stock or squirrelled away in anticipation of breakdowns.

In the period after the Second World War this concept of breakdown maintenance prevailed. It was not until the 1960s that fixed interval overhaul became popular; this entailed maintenance intervention every three months or after producing 50 000 units or running 500 hours or 20 000 miles. The limitation of the regular interval approach is that it assumes that every machine element will perform in a stable, consistent manner. However, in practical situations this does not necessarily apply. There is also the well-known syndrome of trouble after overhaul: a machine which is performing satisfactorily may be disturbed by maintenance work, and some minor variation or defect in reassembly can lead to subsequent problems.

The various approaches to maintenance which followed the fixed interval regime were reviewed in Chapter 1. However, it is interesting to consider some statistics of actual maintenance performance in the early 1990s. Much of the material quoted in the following has been derived from a survey carried out by the journal *Works Management* based on a sample of 407 companies.

Expenditure on maintenance in the EU countries has been estimated at approaching 5% of total turnover with a total annual spend of between £85 billion and £110 billion. This spend is equivalent to the *total* industrial output of Holland, or between 10% and 12% of EU industries' added value. Some 2 000 000 people in 350 000 companies are engaged in maintenance work (Figure 2.1).

| | |
|---|---|
| UK | 5.0% |
| France | 4.0% |
| Italy | 5.1% |
| Spain | 3.6% |
| Ireland | 5.1% |
| Holland | 5.0% |

EU:  £85 billion to £110 billion per year, equivalent to Holland's industrial output
10% to 12% of industries' added value
2 000 000 people in 350 000 companies

**Figure 2.1**  *Maintenance expenditure as a percentage of turnover in EC countries*

- £14 billion annual spend
- Twice UK trade deficit
- 5% of sales turnover
- Three times value of new plant investment
- 18% of book value

**Figure 2.2**  *UK maintenance spending*

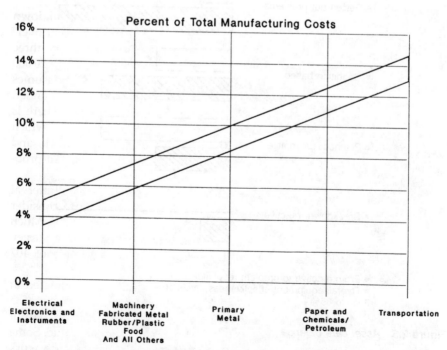

**Figure 2.3**  *Range of maintenance costs by industry (407 sites), UK
Source: Works Management, July 1991*

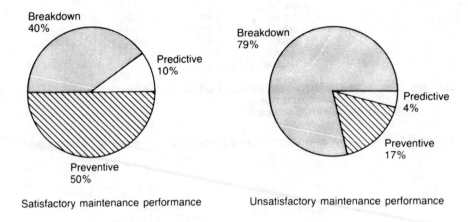

Satisfactory maintenance performance     Unsatisfactory maintenance performance

**Figure 2.4**   *UK type of maintenance*
*Source:* Works Management, *July 1991*

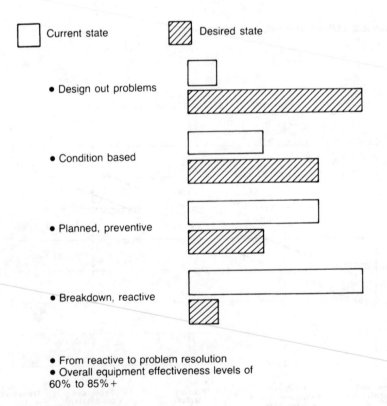

**Figure 2.5**   *Asset care balance*

When we look specifically at the UK, we find the annual spend in 1991 was £14 billion, equivalent to twice the UK trade deficit at that time or 5% of annual turnover. It also equates with three times the annual value of new plant investment in 1991 or 18% of the book value of existing plant (Figure 2.2).

Figure 2.3 gives an indication of the range of maintenance costs in various UK industries expressed as a percentage of total manufacturing costs. The lowest band is around 5% for the electrical, electronic and instrument industries, and the highest averages 12.5% for the transportation industry.

At the time of the *Works Management* survey (1991) the technique most widely employed (40% of companies surveyed) was running inspection. This was followed by oil analysis (27%), on-line diagnosis (25%) and vibration analysis (20%). Fixed cycle maintenance and reliability-centred maintenance came next, but at that time the use of expert systems and the adoption of TPM were both in their infancy, indicating the enormous scope for the application of TPM to UK industry.

Finally, we look at the scope for moving from unsatisfactory to satisfactory maintenance. The pie charts in Figure 2.4 show the potential in moving away from breakdown and towards predictive and preventive approaches. The bar chart in Figure 2.5 serves as an indication of objectives for improvements in performance from the present unsatisfactory levels to benchmark levels.

Japanese methods have been at the heart of the transformation in manufacturing efficiency which has taken place over the last twenty to thirty years and which is still going on. There are common threads running through all of these methods:

- developing human resources
- cleanliness, order and discipline in the workplace
- striving for continuous improvement
- putting the customer first
- getting it right first time, every time.

Central to all these approaches to manufacturing efficiency is the concept of TPM. Asset care has to become an integral part of the total organization so that everyone is aware of and involved in the maintenance function. The end result is that breakdowns become a positive embarrassment and are not allowed to occur. The assets of the production process are operated at optimum efficiency because the signs of deterioration and impending failure are noticed and acted upon.

To achieve this radical change in approach, five principles are involved, and these are covered in detail in Chapter 4. In this context, one of the first and crucial steps towards TPM comes from the application of the five Ss, which are central to all the Japanese methods evolved since the end of the Second World War:

- *seiri*      organization
- *seiton*     orderliness
- *seiso*      cleaning (the act of)

- *seiketsu*     cleanliness (the state of)
- *shitsuke*    discipline (the practice of).

In English-speaking countries an alternative to the five Ss is the more easily remembered CAN-DO of:

- cleanliness
- arrangement
- neatness
- discipline
- order.

The philosophy is exactly the same, however:

1   Get rid of everything and anything unnecessary.
2   Put what you do want in its right place so that it is to hand.
3   Keep it clean and tidy at all times, recognizing that cleanliness is neatness (a clear mind/attitude), is spotting deterioration (through inspection), is putting things right *before* they become catastrophes, is pride in the workplace, giving self-esteem.
4   Pass on that discipline and order to your colleagues so that we *all* strive for a dust-free and dirt-free plant.

The CAN-DO approach therefore is to look at the production facility and clean the workshop and its plant and machinery as it has never been cleaned before, whilst at the same time casting a ruthlessly critical eye at everything in the workplace. Nothing must be allowed to remain anywhere on the shop floor unless it is directly relevant to the current production process. Good housekeeping thereafter becomes everyone's responsibility.

The cleaning process involves the operators of machines and plant. As they clean they will get to know their machines better; they will gradually develop their own ability to see or detect weaknesses and deterioration such as oil leaks, vibration, loose fastenings and unusual noise. As time goes on they will be able to perform essential asset care and some minor maintenance tasks within the limits of their own skills. The process will take place in complete cooperation with maintenance people who will be freed to apply their skills where needed.

With the attitude to cleanliness and good housekeeping understood, we can move on to explain the main principles on which the Japanese approach to TPM is founded. In Chapter 4 we relate these principles to the Western approach and how they are developed in the WCS TPM improvement plan. The main principles of TPM as developed by the Japanese can be described as follows.

### Autonomous maintenance

As operators become more closely involved in getting the very best from their machines they move through seven steps towards autonomous or self-directed maintenance:

1 Initial cleaning.
2 Carrying out counter-measures at the source of problems.
3 Developing and implementing cleaning and lubrication standards.
4 General inspection routines.
5 Autonomous inspection.
6 Organization and tidiness.
7 Full autonomous maintenance.

As these seven steps are taken, over an agreed and achievable timetable, operators will develop straightforward common-sense skills which enable them to play a full part in ensuring optimum availability of machines. At no stage should they attempt work beyond the limits of their skills: their maintenance colleagues are there for that purpose.

## Equipment improvement

The initial process of cleaning and establishing order leads to discovering abnormalities, and progresses through five steps:

1 Discover abnormalities.
2 Treat abnormalities.
3 Set optimal equipment conditions.
4 Maintain optimal equipment conditions.
5 Feed back to design.

The objective of this process is to move progressively towards a situation where all production plant is always available when needed and operating as close as possible to 100% effectiveness. Achieving this goal will certainly not come easily and may take years. The basic concept is one of continuous improvement: 'What is good enough today will not be good enough tomorrow.'

## Quality maintenance

1 Eliminate accelerated deterioration.
2 Eliminate failures.
3 Eliminate defects.
4 Operate profitability.
5 Work as a team.

Everyone is involved in the process – operators, maintainers, team leaders and supervisors. Teamwork is the key to success.

## Maintenance prevention

1 Planned maintenance.
2 Improvements in maintainability.
3 Improvements in work systems and methods.
4 Team working.
5 Aiming at maintenance-free design by feedback to design department.

## Education, training and awareness

1 Establish purpose of training.
2 Establish training objectives.
3 Agree methods of approach.
4 Set up training framework and modules.
5 Design training and awareness programme.

The programme will be designed around the operators and team members concerned. It will be structured to maximize the contribution of each individual and to develop his or her skills to the limit of his or her capability.

In Japan over the last twenty years many hundreds of companies have applied the above principles to their operations. The Japan Institute of Plant Maintenance (JIPM) has carried out stringent audits of TPM achievement and continuity, resulting in the award of PM excellence certificates to successful companies. As already emphasized in Chapter 1, the WCS International approach to TPM is to suitably modify, adapt and apply the Nakajima principles to conditions prevailing in Western countries and industries.

# ─── 3 ───

# *Language of Western TPM*

## 3.1 Setting and quantifying the Vision

Before moving into the necessary detail of the planning, process and measurement of TPM, it is worth while to give an overview of Western TPM and to identify the key building blocks which will be explained in detail and illustrated by case studies in later chapters.

The challenge for many companies is to extend the useful life and efficiency of its manufacturing assets whilst containing operating costs to give a margin which will maximize value to its shareholders and, at the same time, offer enhanced continuity and security of employment. This statement is true whether the particular manufacturing assets are twenty years of age or are just about to be commissioned.

The more forward-thinking companies are linking this challenge to new beliefs and values which are centred on their employees through, for example:

*Integrity*   Openness, trust and respect for all in dealing with any individual or organization.
*Teamwork*   Individuals working together with a common sense of purpose to achieve business objectives.
*Empowerment*   An environment where people are given both the authority and the resources to make sound decisions within established boundaries.
*Knowledge and skills*   Recognizing, valuing and developing the knowledge and skills of their people as a vital resource.
*Ownership*   A willingness on everyone's part to get involved and take responsibility for helping to meet the challenges of the year 2000 and beyond.

Put another way, we can win the challenge by:

- working together
- winning together
- finishing first every time.

This can be delivered by specific values, for example:

- *our people*
- working in a completely *safe* and *fit-for-purpose environment*
- where *quality* is paramount in everything we do
- and where we have a *business understanding* linked to our activities
- and where *reliable equipment*, operated by *empowered* and *effective teams*, will ensure we finish *first every time*.

Total productive maintenance, suitably tailored to the specific environment, can be a fundamental pillar and cornerstone to achieve the above goals, beliefs and values, since:

- We all 'own' the plant and equipment.
- We are therefore responsible for its availability, reliability, condition and performance within a safe and fit-for-purpose environment.
- We will therefore ensure our overall equipment effectiveness ranks as the best in the world.
- We will continuously strive to improve that world class performance.
- We will therefore train, develop, motivate, encourage and equip our people to achieve these goals.

The last statement is the fundamental deliverable if the previous statements are to mean anything in practice.

The development of the maintenance function has been described in Chapter 1. As the aerospace and nuclear power industries, with their relatively complex technologies and systems, emerged in the 1970s and 1980s, we had to respond with a selective and systematic approach. There developed reliability-centred maintenance, which considers the machine or system function and criticality and takes a selective approach, starting with the question: 'What are the consequences of failure of this item for the machine or system, both hidden and obvious?' For example, if the oil warning red light indicator comes on in your car it is obvious you are low on oil. The hidden consequence, if you do not stop immediately and top up the oil, is that the engine will seize! It is therefore good practice to check the oil level via the dipstick at regular intervals. RCM takes a systematic approach using appropriate run-to-failure, planned, preventive and condition-based strategies according to the consequences of failure.

Total productive maintenance uses a similar logic but emphasizes the *people, measurement* and *problem elimination* parts of the equation and not systems alone. It emphasizes that people – operators, maintainers, equipment specifiers, designers and planners – must work as a team if they are to maximize the overall effectiveness of their equipment by actively seeking creative ways and solutions for eliminating waste due to equipment problems. That is, we must resolve equipment-related problems once and for all, and be able to measure that improvement. TPM is a practical application of total quality and empowerment working at the sharp end of the business – on your machines and processes.

Seven useful definitions and benefits of implementing TPM can be summarized as follows:

1    Maximized efficiency of equipment through participation of all employees.
2    Improved reliability of equipment leading to improved product quality and equipment productivity.
3    Economical use of equipment throughout its total service life.
4    Operators trained and equipped to perform minor but essential asset care of their machines.

5   Increased utilization of skilled trades in higher technical areas and more diagnostic work.
6   Practical and effective total quality team working example aimed at equipment improvement and maintenance prevention.
7   Improvement in overall equipment effectiveness as a measurable route to increased profitability.

This seventh definition and benefit is the key: setting the vision is all very well, but we must also quantify that vision and make sure it reflects our business drivers and business objectives. Below is a typical illustration of clear, hard targets for which TPM is the enabling tool.

|  | Benchmark 1990 | Target 8/93 | Actual 8/93 |
|---|---|---|---|
| All lines OEE | 71% | 90% | 88% |
| Model line OEE | 77% | 90% | 92.5% |
| B/D per month | 387 | 40 | 33 |
| Lead time days | 45 | 15 | 15 |
| Accidents | 0 | 0 | 0 |
| Set ups | 16 hrs | 8 hrs | 6 hrs |
| Minor stoppages | 4,650 | 1,000 | 813 |
| Reduction in product costs | 100 | 90 | 91.5 |

We can expand on some of these key definitions and benefits as follows. In TPM, management recognize the simple fact that it is the *operators* of plant and equipment who are in the best position to know the condition of their equipment. Without their cooperation, no effective asset care programme can be developed and implemented. On the contrary they can act as the senses (eyes, ears, nose, mouth, hands) of their maintenance colleagues, and as a team they can work out for themselves the best way of operating and looking after their machines, as well as resolving chronic equipment-related problems once and for all. They can also establish how to measure the resultant improvements.

   TPM involves very little 'rocket science'; it is basically common sense. The problem is, it is quite a rarity to be asked to put our common sense to good use! TPM, however, does just that.


## 3.2   Analogies

In order to illustrate the principles of TPM, three everyday analogies may prove helpful:

- the motor car
- the healthy body
- the soccer team.

Each is described below. At the end of the chapter there is a light-hearted story about an overhead projector operator and his maintenance colleague, which continues the best parts of the analogies in order to underpin the basically simple, but nonetheless fundamental, principles of Western TPM.

### The motor car

A good analogy of using our senses, including common sense, is the way in which we look after our motor cars as a team effort between the operator (you, the owner) and the maintainer (the garage maintenance mechanic).

As the operator of your motor car you take pride of ownership of this important asset. TPM strives to bring that sense of ownership and responsibility to the workplace. To extend the motor car analogy: when you, as the operator, take your car to the garage, the first thing the mechanic will seek is your view as to what is wrong with the car (your machine). He will know that you are best placed to act as his senses – ears, eyes, nose, mouth and common sense. If you say, 'well, I'm not sure, but it smells of petrol and the engine is misfiring at 3000 rpm,' he will probably say, 'that's useful to know, but is there anything else you can tell me?' 'Yes,' you reply, 'I've cleaned the plugs and checked the plug gaps.' He won't be surprised that you have carried out these basic checks, and certainly won't regard them as a mechanic-only job. 'Fine,' he might say 'and that didn't cure the problem?' 'No,' you reply, 'so I adjusted the timing mechanism!' 'Serves you right then,' says the mechanic, 'and now it's going to cost you time and money for me to put it right.' In other words, in the final stage you, the operator, went beyond your level of competence and actually hindered the team effort. TPM is about getting a balanced team effort between operators and maintainers – both experts in their own right, but prepared to cooperate as a team.

As the operator of your car you know it makes good sense to clean it – not because you are neurotic about having a clean car just for the sake of it, but rather because cleaning is inspection, which is spotting deterioration before it becomes catastrophic. The example in Figure 3.1 shows the power of this operator/ownership. In the routine car checks described, our senses of sight, hearing, touch and smell are used to detect signs which may have implications for inconvenience, safety, damage or the need for repairs or replacements. None of the 27 checks listed in the figure requires a spanner or a screwdriver, but 17 of them have implications for safety. The analogy with TPM is clear: failure of the operator to be alert to his machine's condition can inhibit safety and lead to consequential damage, inconvenience, low productivity and high cost.

We don't accept the status quo with our cars because ultimately this costs us money and is inconvenient when problems become major. In other words, we are highly conscious of changes in our cars' conditions and performance using our senses. This is made easier for us by clear instruments and good access to parts which need regular attention. We need to bring this thinking into our workplace.

*Routine checks*
✓ Tyre pressure:    extended life, safety    (eyes)
✓ Oil level:    not red light    (eyes)
● Coolant level:    not red light    (eyes)
● Battery:    not flat battery    (eyes)

Reasons: safety, consequential damage, inconvenience, low productivity, high cost

*Cleaning the car: using our eyes*
● Spot of rust
● Minor scratch
● Minor dent
✓ Tyres wearing unevenly
● Water in exhaust pipe
✓ Worn wipers
● Rubber perishing, trims
✓ Oil leak
✓ Suspension

*Other conditions when operating the machine*
✓ Steering drag    (touch, eyes)
✓ Wheel bearing    (hear)
● Clutch wear    (touch, hear)
✓ Brake wear    (touch, hear)
✓ Exhaust    (hear)
● Engine misfire    (hear, touch)
✓ Engine overheats    (smell)
✓ Petrol leak    (smell)

*One operator to another at traffic lights*
● Exhaust smoke
✓ Front/rear lights
✓ Stop lights
✓ Indicators
✓ Soft tyre
✓ Door not shut

*Message*
No spanner or screwdriver involved in any of the 27 condition checks.

✓ means check has safety implications (17 of 27)

**Figure 3.1**  *Condition appraisal and monitoring: using our senses*

## A healthy body

Figure 3.2 shows our second analogy, which is that healthy equipment is like a healthy body. It is also a team effort between the operator (you) and the maintainer (the doctor).

Looking after equipment falls into three main categories:

*Cleaning and inspection*  The daily prevention or apple a day, which prevents accelerated deterioration or wear and highlights changes in condition. The operator can do most if not all of these tasks.

*Checks and monitoring*  Measure deterioration or use the thermometer, which highlights the trends or changes in performance. The operator can support the maintainer by acting as his ears, eyes, nose, mouth and common sense, so allowing the maintainer to concentrate on the critical diagnostic tasks.

*Preventive maintenance and servicing*  Inject before breakdown, which prevents failure by reacting to changes in condition and performance. The maintainer still does the majority of these tasks under TPM.

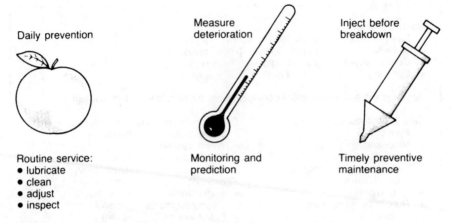

Daily prevention

Measure
deterioration

Inject before
breakdown

Routine service:
• lubricate
• clean
• adjust
• inspect

Monitoring and
prediction

Timely preventive
maintenance

**Figure 3.2**   *Healthy equipment is like a healthy body*

Perhaps the key difference in determining this asset care regime is that under TPM the operator and maintainer determine the routines under each of three categories. If you ask my opinion as an operator or maintainer, and that opinion is embodied in the way we do things for the future, then we will stick with it because it is our idea. On the other hand, if you impose these routines from above, then I might tick a few boxes on a form, but I will not actually do anything!

### The soccer team

The third analogy emphasizes the absolutely crucial aspect of teamwork. At every stage in the development of the TPM process, teamwork and total cooperation without jealousy and without suspicion are essential to success. In Chapter 7 we shall see how these teams are established and developed, but Figure 3.3 gives a pictorial representation of how the teams can function to maximum advantage. In the centre the core team is portrayed: these are the people on the shop floor whose job it is to keep production running at maximum efficiency and minimum losses. Their job is to 'win', just as a soccer team on the field seeks to score and win the match. Just as the soccer team is backed by physiotherapists, coaches and the manager, so the core team has the backing of designers, engineers, quality control, production control, union representatives and management.

In our soccer team the operators are the attackers or forwards, and the maintainers are the defenders. Of course the maintainers can go forward and help the operators score a goal. Similarly the operators can drop back in defence and help stop goals being scored against the team. They are both experts in their respective positions but they are also willing to cooperate, help each other and be versatile. One thing is for sure in the modern world class game: if we do not cooperate, we will most certainly get relegated! The core team will invite functional help on to the shop floor when needed, and all

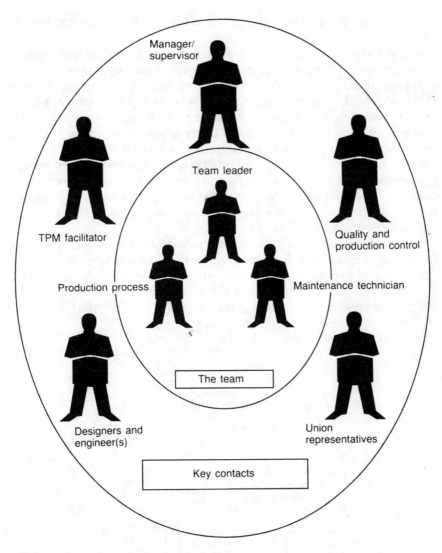

**Figure 3.3**  *Teamworking*

concerned will give total cooperation with the single-minded objective of maximizing equipment effectiveness. Without cooperation and trust, the soccer team will not win. The core team, on the pitch, is only as good as the support it gets from the key contacts who are on the touch line – not up in the grandstand!

The TPM facilitator, or coach, is there to guide and to help the whole process work effectively. People are central to the approach used in Western TPM. We own the assets of the plant and we are therefore responsible for asset management and care. Operators, maintainers, equipment specifiers, designers and planners must work as a team and actively seek creative solutions which will eliminate both waste and equipment-related quality problems once and for all!

## 3.3    Overall equipment effectiveness versus the six big losses

In Figure 3.4 the tip of the iceberg represents the direct costs of maintenance. These are obvious and easy to measure because they appear on a budget and, unfortunately, suffer from some random reductions from time to time. This is a little like the overweight person who looks in the mirror, says he needs to lose weight and does so by cutting off his leg. It is a quick way of losing weight, but not a sensible one! Better to slim down at the waist and under the chin and become leaner and fitter as a result.

The indirect costs or lost opportunity costs of ineffective and inadequate maintenance tend to be harder to measure because they are less obvious at first sight – they are the hidden part of the iceberg. Yet they all work against and negate the principles of achieving world class levels of overall equipment effectiveness, or OEE as it is called.

In our iceberg example the impact on profitability is in inverse proportion to the ease of measurement. Quite often we find that a 10% reduction in the direct costs of maintenance (a commendable and worthwhile objective) is equivalent to a 1% improvement in the overall effectiveness of equipment, which comes about from attacking the losses that currently lurk below the

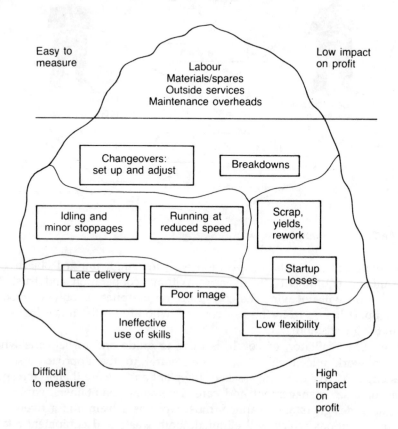

**Figure 3.4**    *True cost of maintenance: seven-eighths hidden*

surface. Fortunately the measurement of cost/benefit and value for money are central to the TPM philosophy.

In most manufacturing and process environments these indirect or lost opportunity costs include the following, which we call the *six big losses*:

- breakdowns and unplanned plant shutdown losses
- excessive setups, changeovers and adjustments (because 'we are not organized')
- idling and minor stoppages (not breakdowns, but requiring the attention of the operator)
- running at reduced speed (because the equipment 'is not quite right')
- startup losses (due to breakdowns and minor stoppages before the process stabilizes)
- quality defects, scrap and rework (because the equipment 'is not quite right')

In addition to the six losses we create a situation characterized by

- late delivery to our customers
- a poor image for ourselves and our customers
- ineffective use of our inherent skills
- low flexibility to respond to our customers and our problems!

In Figure 3.5 we show the six big losses and how they impact on equipment effectiveness. The first two categories affect availability; the second two affect performance rate when running; and the final two affect the quality rate of the product. What is certain is that all six losses act against the achievement of a high overall equipment effectiveness.

In promoting TPM equipment improvement activities you need to establish the overall equipment effectiveness rate as the measure of improvement. The OEE formula is simple but effective:

$$\text{OEE} = \text{availability} \times \text{performance rate} \times \text{quality rate}$$

You will also need to determine your ultimate world class goal or benchmark on the OEE measure. This should not be an idle dream: rather it should be realistic, exacting, demanding and better than your competitors. Take each of the three elements in turn and set your ultimate goal. There may

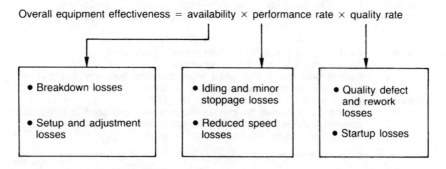

**Figure 3.5** *Measuring equipment effectiveness*

be a strong argument for saying that the performance rate of the plant should be nothing less than 100%, but be realistic. Figure 3.6 shows an example which happens to give an OEE of 85%: your target may be higher.

Start to run the three measures, week by week, on your critical machines, lines and processes. Build up the notion of the 'best of the best'. It is a very powerful and strong case. If we take the example shown in Figure 3.6, the best availability (week 2) times the best performance rate (week 1) times the best quality rate (week 4) gives an OEE of 80%. What stops you achieving the best of the best consistently? The answer is that you are not in control of the six losses mentioned above. Get control, gradually reduce and then eliminate the six losses through TPM, and you will become world class. This best of the best has a high belief level: 'We have achieved it at least once in the last $x$ weeks; the problem is we do not achieve each of the three OEE elements consistently.'

Each 1% improvement on the OEE represents a significant contribution to profitability: it is the improvement below the tip of the iceberg. The vital issue is, of course, to determine what you can do with the improvement. Let us take a simple example:

|  | OEE | Good units produced | Time taken |
|---|---|---|---|
| Current | 60% | 1000 | 80 hours |
| Best of best | 75% | 1250 | 80 hours |
|  |  | or  1000 | 64 hours |

In the above example, consistent achievement of the best of the best OEE from a 60% base to 75% is a 25% improvement. This means you can either produce 250 more units in the same time *or* the same number of units in 25%

| Overall equipment effectiveness (%) | | = | availability of plant (%) | × | performance rate of plant (%) | × | quality rate (%) |
|---|---|---|---|---|---|---|---|
| Ultimate goal (world class) | 85 | = | 90 | × | 95 | × | 99 |
| Actual:  week 1 | 75 | = | 85 | × | 93 (best) | × | 95 |
| week 2 | 76 | = | 88 (best) | × | 90 | × | 96 |
| week 3 | 72 | = | 86 | × | 91 | × | 92 |
| week 4 | 68 | = | 82 | × | 85 | × | 98 (best) |
| Average | 73 | = | 85 | × | 90 | × | 95 |
| Best of best (target) | 80 | = | 88 | × | 93 | × | 98 |

| Question: | What is stopping us achieving best of best consistently? |
|---|---|
| Answer: | We are not in control of the six big losses |
| However: | Best of best has a high belief level; therefore teamwork and problem solving will lead to elimination of the six losses. |
| Question: | What is each 1% improvement worth on OEE? |
| Answer: | It depends of course on the specific company and plant situation; but experience shows that a 1% improvement in OEE is often equivalent to a 10% reduction in the direct cost of maintenance. |

**Figure 3.6**  *TPM performance measure*

less time – or, of course, some combination between these two levels. The key point is that consistent improvement in the OEE gives the company and its management *a choice of flexibility which they do not currently enjoy* at the 60% OEE level.

Figure 3.7 presents most of the previous points as a summary of TPM's most desirable effects and the resultant benefits. TPM also gives us a clear vision, direction, involvement, empowerment and measurement tool for our future overall equipment effectiveness.

## 3.4 Getting started in your plant

As with most good practices, there is nothing particularly earth-shattering about TPM. The essence lies in the ability to focus the concepts and principles on the reality of the actual day-to-day situation. This means getting the climate right through front-line teamwork, aiming for motivation and ownership of the condition and productivity of equipment when it is up and running, rather than the 'I operate, you fix' traditional approach. This is easily said, but is potentially difficult to implement unless TPM is both Westernized and tailored to the specific industry and local plant environment and business drivers (uptime versus downtime, output production, maintenance cost per unit of output, safety considerations, job flexibilities and so on) and of course the *essential* cultural and attitudinal perspectives.

Essentially we are talking about new ways of working, more effective and cooperative methods of doing essential asset care tasks and equipment-related problem resolution. This is achieved by improving the flexibility and interaction of maintenance and production, supported by excellent

| Feature | Result |
| --- | --- |
| ● Machines run close to name-plate capacity | ● Reduced capital expenditure need |
| ● Ideas to improve often proposed by operators | ● Ownership/success |
| ● Breakdowns rare, and we achieve flawless operation | ● Used to learn and teach the team |
| ● Machines adapted to our need by our people | ● Our machines will be better |
| ● Operators and maintainers solve problems themselves | ● Fewer delays and stoppages: enhanced self-esteem |
| ● Cleanliness and pride in continuous improvement | ● Good working environment |
| ● More output potential from existing plant | ● More profits and/or more control and choice |

**Figure 3.7** *TPM vision of the future*

management, supervision, engineers and designers, plus systems, documentation, procedures, training, quality and team leading within an environment where safety is paramount.

TPM experiences in a wide range of industries confirm that it is essential to put handles on the issues before you can start to formulate a realistic programme of TPM-driven improvement with associated training, awareness and development. There is only one way to put handles on the issues, and that is to see and feel them at first hand. You *must* be prepared to spend sufficient time in the selected or proposed TPM plant so you can see the reality and talk to the managers, superintendents, supervisors, engineers, designers, technicians, craftsmen and operators. As a result, you can understand where the plant is today and where it can realistically go for the future using the TPM approach. Whilst in the plant you can also formulate the training and awareness requirements as a properly thought-out plan with clearly identified benefits, costs, priorities, milestones, timescales, methods and resources. Each and every plant is like a thumb-print, it is unique and has to be treated as such.

## 3.5   TPM implementation route

In helping our customers to introduce TPM principles, philosophy and practicalities into their company, we have developed a unique and structured step-by-step approach which is illustrated in Figure 3.8, and is based on the following:

1   An initial half-day to one-day general TPM awareness session for senior management, supervision and employee representatives. This is the essential first step to get 'buy-in' and initial commitment from the key influencers.
2   A short, sharp scoping study to determine how TPM can be tailored to suit the specific plant situation, typically based on initial TPM pilots linked to a site roll-out programme.
3   A mobilization phase to agree to commitment, delivery, site awareness and detailed training for the TPM pilots and subsequent site-wide implementation.
4   Plant-specific TPM awareness module for all employees directly or indirectly involved in TPM.
5   A comprehensive and practical four-day TPM facilitator and supervising training course.
6   Launching and supporting the TPM pilots including comprehensive TPM awareness sessions and training of management and support staff, team leaders and team members including extensive and practical on-the-job coaching.
7   Parallel corporate and plant-level project management support and reviews to ensure a coherent and consistent TPM policy, methods, publicity, standards and measurement process with clear completion criteria.

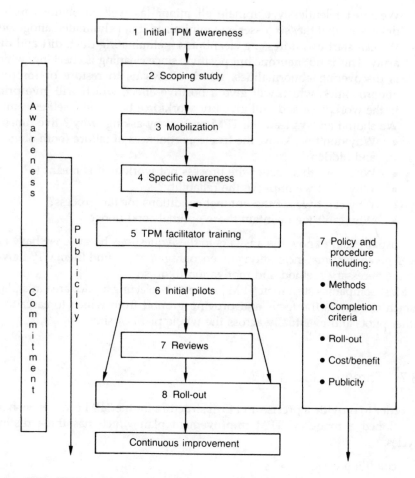

**Figure 3.8** *TPM implementation route*

8 Site-wide roll-out planning and implementation based on a continuous improvement culture.

In our experience it is *vital* to tailor your TPM implementation plan, not only to suit the differing Western cultures and industry types, but also to recognize the sensitivity of local plant-specific issues and conditions.

## 3.6 What is the 'on-the-job' reality of TPM?

TPM is different from other schemes. It is based on some fundamental but basically simple common-sense ideals:

1 We must restore equipment before we can improve its overall effectiveness.
2 We can then pursue ideal conditions.

3   We must relentlessly eliminate all minor (as well as obvious major) defects, so that the 'six losses' are minimized if not eliminated altogether.
4   We can start by addressing cleanliness – eliminating dust, dirt and disarray. This is not neurotic but positive, since cleaning is checking, which is discovering abnormalities, which allows us to restore or improve abnormalities, which will give a positive effect, which will give pride in the workplace and will give our workforce back some self-esteem.
5   We should always lead the TPM process by asking 'why'? five times:
    - Why don't we know the true consequences of failure (both obvious and hidden)?
    - Why does this part of the process not work as it is meant to?
    - Why can't we improve the reliability?
    - Why can't we set the optimal conditions for the process?
    - Why can't we maintain those optimal conditions?

We usually don't know the answers to these questions because we have not been given the time, inclination and encouragement to find them. TPM gives us the necessary method and motivation to do so.

Most companies start their TPM journey by selecting a pilot area in a plant which can act as the focus and proving ground from which to cascade to other pilots and eventually across the whole plant or site.

## 3.7   Three cycles in TPM

In order to provide a precise and firm structure for the TPM process we have developed a nine-step TPM improvement plan which has three distinct cycles:

- condition cycle
- measurement cycle
- improvement cycle.

For each TPM pilot area or equivalent, the TPM team members are taken step by step through the main elements as shown in Figures 3.9 and 3.10.

The following three cycles and nine steps are fully developed in chapter 5.

### Condition cycle

### Criticality assessment

In order to decide which are the most critical plant items, the TPM equipment team list the main assets. Then they independently assess each of the assets from their perspective, and rank them on a scale of 1 (low) to 3 (high) regarding criteria such as maintainability, reliability, impact on production quality, sensitivity to changeovers, knock-on effect, impact on throughput velocity, safety, environment and cost. The team should reach a consensus of ranking and weighting on the most critical items.

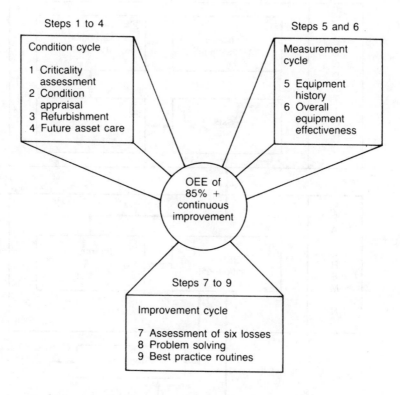

**Figure 3.9** *TPM improvement plan: three phases and nine steps*

## Condition appraisal

Following the above step, the TPM pilot teams can start the equipment condition appraisal. Typically a team will comprise team leader with operators and maintainers as team members plus of course the TPM facilitator. The TPM pilot teams should also be supported by their key contacts: this would typically comprise management/supervision, quality engineers, safety engineers, production, process, industrial engineers and designers, whom the team can call in to help resolve specific equipment-related problems.

## Refurbishment

The next task of the TPM team following appraisal of the condition of the pilot equipment is to decide what *refurbishment programme* is required to restore the equipment to an acceptable level of condition from which the TPM team can pursue the ideal conditions.

THE TPM IMPROVEMENT PLAN

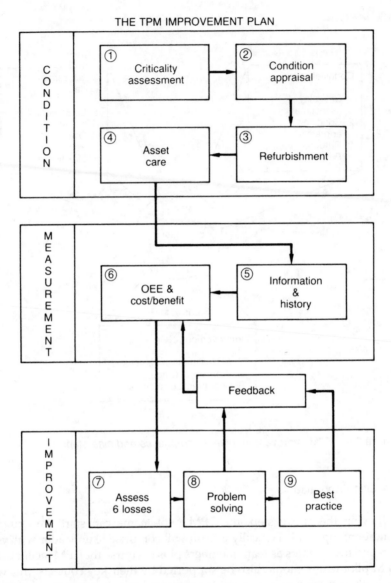

**Figure 3.10**   *TPM improvement plan sequence*

## Future asset care

Whilst completing the condition appraisal, the team can also determine the future asset care programme in terms of who does what and when. This includes the team deciding the daily prevention routines, the lubricate, clean, adjust, inspect activities, most of which can be done by the operations staff. The team also decide the condition monitoring activities to measure deterioration, remembering that the best condition monitor is the operator who acts as the ears, eyes, nose, mouth and common sense of his maintenance colleague and can call him in when things start to go wrong and before they become catastrophic. Finally the team decide on the regular PPMs – the planned, preventive maintenance routines. Under full TPM the operator performs most of the daily routines; the operator and the maintainer both contribute to condition monitoring; and the maintainer does the PPM scheduled work.

## *Measurement cycle*

### Equipment history record

The TPM team will also determine the records to be kept with regard to the history of the equipment which will aid in future problem resolution.

### OEE Measurement and potential

In parallel with this exercise the team carry out the initial measurement of overall equipment effectiveness in order to determine current levels of performance, the best of the best interim targets and the ultimate world class levels.

## *Improvement cycle*

### Assess 6 losses

Assessment and scope of the potential for eliminating the six losses through P-M analysis

### Problem solving

P-M analysis is a problem solving approach to improving equipment effectiveness which says: there are *phenomena* which are *physical*, which cause *problems* which can be *prevented* (the four Ps), because they are to do with *materials, machines, mechanisms* and *manpower* (the four Ms). This is the acid test of the TPM pilot(s) since the teams are trained, encouraged and motivated to resolve (once and for all) the six losses which work against the achievement of world class levels of overall equipment effectiveness. These problem solving opportunities can usually be classified as: operational problems/improvements involving no cost or low cost, and low risk; technical problems/improvement often involving the key contacts and also some cost and hence risk; and finally support services and/or support equipment problems/improvements which again quite often involve the key contacts and some low cost and low risk.

Best practice routines

Finally, the TPM pilot team will develop its own BPRs regarding the equipment operation and asset care policy and practice. All these feed back into an improved OEE score which will encourage the continuous improvement 'habit' which is central to the TPM philosophy. As in total quality, the personnel will also become empowered!

In order to summarize the previous explanation, Figure 3.11 shows that TPM involves a team of craftsmen and operators who are supported by their key contacts and who follow the TPM improvement plan through initial pilots in order to eliminate the six major losses. Their progress is measured by improvements in the overall equipment effectiveness which allows the team to understand the need to continuously improve. Finally, the TPM

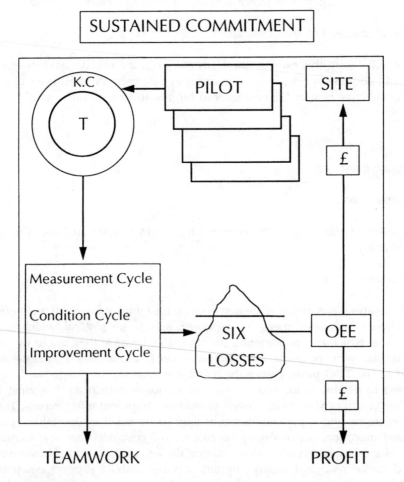

**Figure 3.11** *TPM cycle*

process will only work provided it has the sustained commitment of everybody – which, of course, must start from the top.

## 3.8 Win commandments for successful TPM implementation

The ten essential rules or win commandments for the successful implementation of TPM can be stated as follows:

1 *Practical* Recognize TPM as a practical application of total quality and hence world class performance aimed at improving the overall effectiveness of our equipment through *people* and effective teamwork, rather than just systems.
2 *Commitment* There must be 100% commitment from the top of the organization linked to total sustained consistency in the application of the other nine commandments.
3 *Milestones* Set three- to five-year milestones showing how it is intended to implement TPM as part of the company's vision of the future. It is not a programme which has a starting and finishing point; it is a process, a process of continuous improvement.
4 *First-year objectives* From the strategic milestones, set a detailed one-year programme with clearly defined identified activities, roles and responsibilities, all aimed at getting the TPM process started and probably focusing on a pilot project (see rule 9).
5 *Attention to detail* TPM demands 100% attention to detail. It is the small, difficult to identify, apparently minor equipment malfunctions and poor or irregular equipment operation which prevent the achievement of world class performance. The big events, such as breakdowns, are obvious and hence can be tackled with vigour. The winning car in Grand Prix racing succeeds because of a team effort, with 100% attention to detail to ensure every extra millisecond of pace and precision of handling.
6 *Overall equipment effectiveness* Decide how OEE is to be applied and measured. Decide how you will make use of the improvements.
7 *World class* Achievement of world class standards must be based on OEE using the notion of best of the best (BOB) to set interim targets. Because the three elements of BOB have actually been achieved at least once over a reference period, BOB can be seen to be achievable. Success will come from the regular achievement of BOB once the six big losses have been identified, controlled and then eliminated.
8 *Training* Improvements in equipment performance and people's attitudes and behaviour will not be achieved without training. The benefits of training are not immediately obvious but, as the saying goes, 'If you think training is expensive, try ignorance: that can be really costly!' With TPM the awareness and training journey must begin – from innocence to excellence. Supervisors and middle managers must be involved in the training process; they are often the most difficult to convince that empowerment of the people is a good thing.

9    *The pilot project*   Select an area for a pilot project, preferably a notorious trouble-spot where the benefits can easily be measured and will be obvious to all. The project will be most effective in an area where attitudes are also already reasonably positive.

10   *Do not expect miracles*   Practical measurable results will begin to flow from Day One, but patience will be needed. An environment is being created for people to take ownership of their plant and equipment. They will need to change their way of working and especially their way of thinking about their work. This is not an easy process for the whole company, and particularly not for operators and maintainers. 'If you are patient,' as Taiichi Ohno, President of Toyota, once said, 'you will be rewarded many times over by your people. They do have a mind of gold; they do have self-esteem.' TPM will tap that mind of gold.

It must be appreciated that the ten win commandments are not hard and fast but must be adapted to the organization and its people. An example of adopting the above to suit local circumstances is shown in Figure 3.12.

## 3.9   Attractions, benefits and advantages of TPM

In a single sentence, TPM is driven by *business benefits* but is a *grass roots* process, which enables *communication* with the shop floor in a structured and coherent way and leads to *total quality* with *measurable results*. It is about achieving customer satisfaction, cost competitiveness and world class performance (Figure 3.13). It is based on five beliefs:

- people
- teamwork
- continuous innovation and improvement
- customer satisfaction
- winning and prosperity

and it is delivered to support:

- lean production
- just in time

### TPM attractions

- It is measurable, visible, practical common sense.
- Employees can understand and therefore value the concept.
- The elements of TPM have been tried and tested.
- Much of TPM is about rediscovering old values.
- It is a grass roots process.
- It is also a world class process
- It is led by manufacturing and therefore driven by production and maintenance as equal partners.

TPM win commandments

1 We need the sustained and active commitment from corporate boardroom to shop floor.
2 We must always pursue the CAN-DO approach and ensure that each piece of equipment is seen as a building block in the overall structure of the unit and its effectiveness.
3 We should use the improvement of the OEE to promote a sense of achievement, pride, and individual and team ownership.
4 We should increase the knowledge of the TPM process within the team environment and use this to develop and meet training requirements.
5 We should aim to eliminate the six major losses and show the reasons why, in order to promote the value of TPM.
6 We should use the TPM process to promote and improve teamwork across all functions.
7 We should standardize all equipment performance measurement to make it easy to understand and use.
8 We should set achievable short-, medium- and long-term goals within a realistic plan.
9 We should develop belief in TPM using recorded, proven and visible results communicated to all.
10 We will put all our equipment under a permanent microscope.

**Figure 3.12** *TPM win commandments*

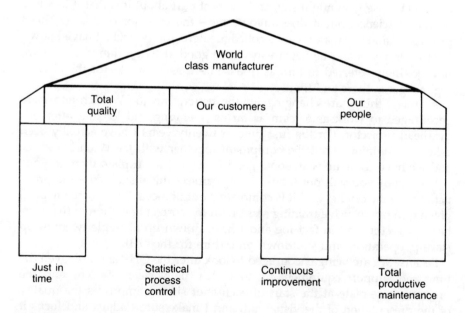

**Figure 3.13** *Achievement of world class performance*

- It is about improving equipment effectiveness through people and not through systems alone.

### Lessons learned from TPM

- It focuses on the lost opportunity costs of ineffective maintenance.
- The expression 'It must be OK if it is up and running' is no longer valid.
- Skills are enhanced, not diluted.
- Craftsmen become the doctors and teachers of machine health care.
- TPM establishes the habit of continuous improvement.
- TPM encourages the outlook: 'If it's my idea, I'll stick with it.'

## 3.10   Overhead projector analogy

Good morning everybody. My name is Peter Willmott and my job is to operate this overhead projector (Figure 3.14). I have worked for the OHP Company for twenty years now and, provided I have walked in vertically every day and have been warm to the touch, nobody in management has taken too much notice of me!

Times are a-changing, however, and seemingly for the better. Apparently our managing director has been to visit our competitors in Japan and has seen how they look after their equipment and, perhaps more importantly, their people. They practise a thing called total productive maintenance or TPM. Now that it has been explained to us, we prefer to think of TPM as teamwork between production and maintenance. It's quite simple really. Common sense, you might say, and the best part about it is that it involves no rocket science; but it does involve me – the operator of this overhead projector – and my maintenance colleague here, Joe Wrench. I have known Joe for ten years and he has always been good at fixing things. In fact we have jokingly referred to him as 'Joe'll Fix-It' as he works for the GITAFI regime: get in there and fix it!

As I say, things are changing for the better, and Joe Wrench and I are encouraged to work as a team as far as operating and looking after this overhead projector. For the first time in twenty years I have actually been asked my opinion about the equipment together with Joe Wrench's ideas, and we have come up with some good ideas. Let me explain them to you.

For a start you will not thank me if I project this visual aid – the product – on to the ceiling, or if it is completely out of focus. I have actually been given a comprehensive training session on the correct operation of the overhead projector, and in fact Joe and I have drawn up a simple twenty-step startup, operation, and shutdown procedure for the OHP.

Because we are being encouraged to look after the OHP and are given the time and support equipment to do it, I actually clean the lens and the projector base-plate at the start of each shift since it improves the quality of the presentation of the visual aid, and I make sure I adjust and focus it correctly before starting the shift. By the way, I also make sure I cover up the

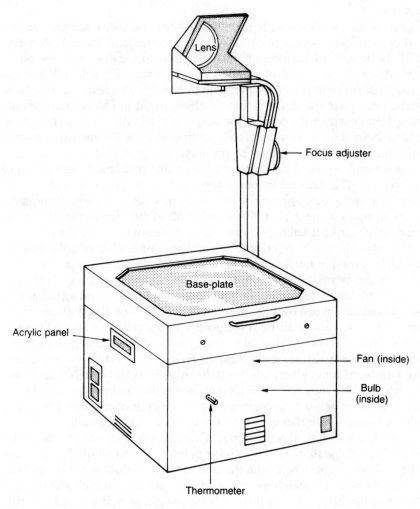

**Figure 3.14** *Overhead projector*

base-plate of the OHP at the end of the shift as it can easily get scratched and damaged if I do not do this simple chore. A new base-plate for this OHP costs £35.00, which is about 15% of the cost of a complete new OHP. It is also inconvenient as it takes about three hours to change over the bad one for a new one.

Anyway, as I said, Joe Wrench my maintenance colleague and I have been given the training, time and encouragement to sort out the best way of running this piece of equipment. Let me tell you what *we* have decided to do. Not, you will notice, some clever chap from central planning, but Joe and I. We are in a team now, and Fred Whitlock the ex-supervisor is our team leader. Since he has been on a facilitator course, he's changed for the better: he actually asks our opinion about things and he actually takes time out to come down here to listen and discuss better ways of doing things with Joe and me.

One of the problems with this OHP is that the focus adjuster on this vertical arm here seems to wear out quite often, and the ratchet won't hold the lens head in focus. If this happens during the shift, Joe and I have decided that we don't need to actually stop the shift for a major repair. Instead I can pin the ratchet with this wedge as a temporary measure whilst I complete the shift. We can, in effect, run it to failure, and the only thing I make sure I do is to let Joe know that he will need to change the ratchet focus adjuster as soon as he's got time to do it. By the way, Joe and I have put forward a proposal to production engineering to use a closer tolerance and age-hardened ratchet so that this problem is resolved once and for all. This will mean our spares costs will go down and I won't have to mess about jamming a wedge in here as a temporary measure: botching up is a thing of the past. After all, if the handbrake ratchet on your motor car kept failing, you wouldn't put up with it, would you?

The other thing which Joe and I have discussed is changing the bulb on this OHP. I used to think it was the most critical part of this machine, but it isn't. It's important, but not as critical as something else which I will tell you about later. Anyway, back to the bulb. They do fail now and again, and in the bad old days when the bulb went I used to switch off the machine and go for a cup of tea and wait until Joe got round to getting a new one out of stores – which is about half a mile away. Joe would then change it over and we would eventually get going again. I reckon we used to lose something like four hours a month on this 'breakdown' if the bulb went. Not anymore, though, because Joe and I have thought about this problem as well. In fact I've been on a half-day in-company bulb changing course and I'm now a fully accredited bulb changer – and I'm certainly no electrician.

What happens now if a bulb goes? Quite simple: I switch off the on/off switch here and walk over to the power point. I switch that off. I pull out the plug and bring it back here with the lead, so there is absolutely no way I can electrocute myself. I then remove the lid, take out the old bulb and put it in the waste-bin here; I do not leave it lying around as a future accident risk. Then, using a cloth, I take out the new bulb from its packaging; I use a cloth because it's a halogen bulb and if I get my sweaty hands on it, it will be useless. I then insert it here, replace the lid, take the plug and lead to the socket and re-energize the circuit. Switch on the on/off switch at the OHP and 'bingo' – we're back in production.

There are some other important points about bulb-changing which Joe and I have agreed. We keep two spare bulbs here on the machine – not 800 metres away in central stores. I always – without fail – record the fact that I've used a bulb for two reasons: first to get a spare replacement organized from central stores; and secondly so that Joe and I can build up an equipment history file on this OHP so we both have access to past problems – non-standard events, if you like – which will help us on our problem resolution sessions. At the moment Joe and I are looking into the possibility of bulbs with a different power rating as these current ones seem to be unreliable. Joe also thinks it may be something to do with dust and dirt ingress, but more on that later. The way we've organized bulb changing is part of our planned preventive maintenance (PPM) schedule for this OHP.

Now to the best part of this equipment care procedure which Joe and I have built up and which we are both pretty proud of. The most critical part of this OHP – given that we have a power supply of course! – is the fan. If the fan goes, the bulb will most certainly blow, and my product, the overhead view foil, will probably melt in the process! I hadn't really thought about this before – mainly because I hadn't been asked to think about it!

It's quite interesting, really, because when I started to clean the OHP I could tell if it was getting hot – or overheating, to be precise (because I like to be precise nowadays). Cleaning is inspection, and I'm really acting as the ears (sound), eyes (seeing), nose (smelling), mouth (taste), hands (touch, heat, vibration) and *common sense* of my maintenance colleague Joe Wrench. And I don't usually need a spanner or screwdriver to use any of my God-given senses! Incidentally, I've learnt that common sense is in fact quite uncommon unless we're encouraged to use it! Anyway, back to the fan – the most critical part of my machine, the OHP. If, when I'm cleaning it, I notice it's getting too hot, or if it starts to make a noise or vibrates, I do one thing, and one thing only. I switch off the on/off switch, I pull out the plug and bring it back here to the machine, and then I get Joe to come and see what's wrong. It's beyond my level of competence or skill at the moment to go messing about with the fan, but I can and do act as the early warning system for Joe.

In fact Joe and I have thought a lot about the fan and we are getting a bit more scientific about the early warning system – or condition-based maintenance or monitoring (CBM) as we call it. Rather than trust my 'feeling the heat' or 'hearing the noise' senses we've decided to drill a hole here in this precise position and we've inserted a thermometer with a red mark on the 40 °C point. So, during the shift I do three readings: after one hour, after four hours, and just before the end of the eight-hour shift. I can trend the readings and I keep them up on this visible wall chart so that both Joe and I can see the temperature trends, alongside the major event fault trends. Joe and I have made two other improvements as well. In fact we're quite proud of these equipment improvements that we've implemented. The fan drive belts used to break quite often, so I suggested we cut out a 50 mm × 50 mm panel on the side here and put an acrylic cover in place of the metal sheet we removed, so that we can look inside the OHP base and see if the belt is fraying before it actually breaks. It's simple, really, and we think quite effective. In fact all 80 of our other OHP machines are now fitted with the thermometer and the acrylic cover modifications.

Joe and I have developed best practice routines (BPRs) for this OHP, and they divide into three main areas:

- the 'apple a day' routines which I do as a matter of habit
- the 'thermometer' or condition monitoring routines which Joe and I share
- the 'injection needle' or planned maintenance which is mainly carried out by Joe.

The point is that *we*, not someone from on high, have decided the best practice routines to operate and take care of *our* asset – the OHP. Also *we* have decided *who* actually carries out each asset care task, *how* we carry it out, with *what* frequency and with *what* support tools and equipment.

It's our ideas, it's our disciplines that are important: we've got *ownership* and we work as a *team*. We've been given the *time*, the *responsibility* and the necessary *training* and *encouragement* to take *ownership*, and *we like it*. It's given us back some *self-esteem*. It's for maintenance to be productive, whoever does it!

Finally, Joe and I had our photographs put up in the reception area in the front office last month as recipients of the 'TPM team of the month' award. Silly, really, but Joe and I felt quite good about it. Even my wife says I'm warm to the touch now!

# 4

# Relationship between Japanese and Western TPM

The key significance of Nakajima's work in the evolution of TPM and the differences between the work ethic in Japan and that in the West have already been referred to in Chapter 1. Nakajima established five pillars for the application of TPM:

1  Adopt improvement activities designed to increase the overall equipment effectiveness (OEE) by attacking the six losses.
2  Improve existing planned and predictive maintenance systems.
3  Establish a level of self-maintenance and cleaning carried out by highly trained operators.
4  Increase the skills and motivation of operators and engineers by individual and group development.
5  Initiate maintenance prevention techniques including improved design and procurement.

One of the main purposes of this book is to show how Nakajima's pillars can be adapted to meet Western needs. To reiterate the analogy in Chapter 1: 'In a heart transplant operation, if you do not match the donor's heart to that of the recipient, you will get rejection.'

Seiici Nakajima's answer to the question 'What is TPM?' provides at least three basic aims:

- TPM aims to double productivity, and reduce chronic losses to zero.
- TPM aims to create a bright, clean and pleasant factory.
- TPM aims to reinforce people (empower) and facilities and, through them, the organization itself.

The differences are not just between Japan and the West; the approach required will vary from country to country, from industry to industry and from one company to another. Experience has shown that tailoring TPM to the local, plant-level organization and its people is the only way to achieve success. This process must be founded on wide experience of applying TPM in different countries and in different industries, whilst at the same time recognizing local, plant-specific issues.

As explained in Chapter 3, Western TPM identifies three cycles in applying the pillars or principles:

- condition cycle
- measurement cycle
- improvement cycle.

The present condition and future asset care requirements for the plant and equipment are first established and then developed through the measurement cycle, which sets the present and future levels of overall equipment effectiveness. Finally, the improvement cycle carries the process forward to the best of best and on to world class through a continuous improvement 'habit' (this concept is fully developed in Chapter 6).

Figure 4.1 shows the relationship between Nakajima's approach and that of WCS International. Figure 4.2 shows the driving force behind Japanese TPM, and Figure 4.3 shows the approach pioneered by the Japanese Institute of Plant Maintenance.

The scope for improving on the way we do things now can only be established by adopting the continuous improvement approach and by never accepting that what we are achieving today will be good enough for the future. A striking example of this comes from a visit by the author some years ago to the press shop in a Toyota automobile plant in Japan, where it was observed that a 1,500 tonne press die change took place in the astonishingly short time of $6\frac{1}{2}$ minutes. When this was commented on, the reply came: 'Yes, yes, we know, we need to reduce the time to 5 minutes.' At that time a comparable change in a UK plant could take up to 4 hours, with an improvement objective of 2 hours. In today's Japanese car plants a straightforward die change is regularly achieved in a single minute!

Analogies and visual aids are essential components in the process of introducing TPM. One of these is the concept of healthy equipment, as already illustrated in Figure 3.2, which portrays the 'apple a day' for good health, the 'thermometer' to monitor well-being and the 'injection' to protect against disease. Routine asset care involving lubricating, cleaning, adjusting

| | Condition cycle | Measurement cycle | Improvement cycle |
|---|---|---|---|
| 1  Attack six losses: improve OEE | ✓ | ✓ | ✓ |
| 2  Set up planned, preventive maintenance | ✓ | — | — |
| 3  Establish autonomous maintenance | ✓ | ✓ | ✓ |
| 4  Training and education | ✓ | ✓ | ✓ |
| 5  Equipment improvement, maintenance prevention | — | ✓ | ✓ |

**Figure 4.1**  *Relationship between five pillars (Nakajima) and Western TPM improvement plan*

| | |
|---|---|
| Goal: | economic world domination via |
| Flexibility: | right products |
| | right time |
| | right quality |
| | right price |
| Trouble free: | zero defects |
| | zero equipment failures |
| | zero accidents |
| Stockless: | no buffer stocks |
| | no WIP |
| | |
| All equals: | total waste elimination |

TPM viewed as an essential pillar for equipment reliability and product repeatability through people and not systems alone

**Figure 4.2** *Essence of Japanese TPM*

Total productive maintenance (TPM) combines the conventional practice of preventive maintenance with the concept of total employee involvement. The result is an innovative system for equipment maintenance that optimizes effectiveness, eliminates breakdowns and promotes autonomous operator maintenance through day-to-day activities.
 Specifically, TPM aims at:

1 establishing a company structure that will maximize production system effectiveness
2 putting together a practical shop floor system to prevent losses before they occur, throughout the entire production system's life cycle, with a view to achieving zero accidents, zero defects and zero breakdowns
3 involving all departments including production, development, sales and management
4 involving every single employee, from top management to front-line workers
5 achieving zero losses through small-group activities.

**Figure 4.3** *What is TPM? The JIPM definition*

and inspecting ensures the plant is protected against deterioration and that small warning signs are acted upon. Condition monitoring and prediction of impending trouble ensure that developing minor faults are never allowed to deteriorate to a breakdown or a reduced level of machine effectiveness. Finally, timely preventive maintenance safeguards against the losses which can come from breakdowns or unplanned stoppages.

 Let us now take a closer look at each of the five Nakajima TPM principles together with the condition, measurement and improvement cycles of the TPM improvement plan (Figure 4.1).

## 4.1 First principle: attack the six losses to improve OEE

Figure 4.4 illustrates how the OEE links to the six losses, and demonstrates that central to the philosophy of TPM is the identification of reasons for, causes of and effects of the six losses, such that their elimination is bound to lead to an improvement in the OEE. An example from the offshore oil industry shown in Figure 4.5 illustrates that poor asset care can lead to

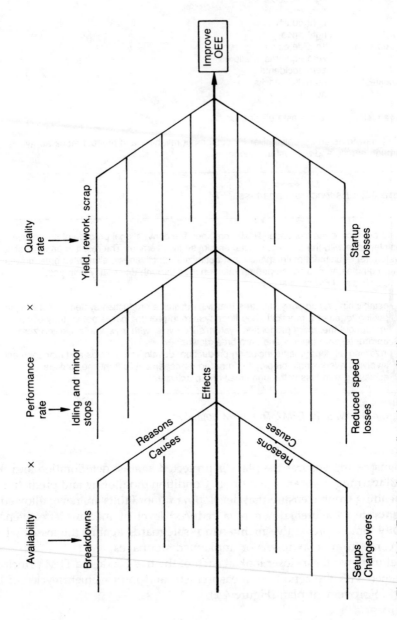

**Figure 4.4** *Factors in overall equipment effectiveness*

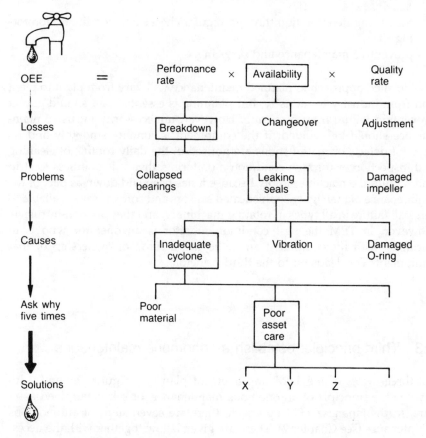

**Figure 4.5** *Problem solving cascade*

inadequate cyclone operation, which gives rise to a leaking seal problem, which results in a breakdown, which affects the availability part of the OEE measure.

In the TPM improvement plan (Figures 3.9 and 3.10) these aspects are covered in the condition, measurement and improvement cycles through steps 4 to 9.

## 4.2 Second principle: set up planned, preventive maintenance

In the TPM improvement plan (Figures 3.9 and 3.10) this aspect is covered by steps 1 to 4 of the condition cycle. The key step is number 4, where the future asset care regime is determined by the TPM team, and is based on the principles illustrated in Figure 3.2:

• daily prevention, cleaning and inspection

- measuring deterioration through regular checks and condition monitoring
- preventive maintenance and servicing.

The detailed approach to planned maintenance will vary from plant to plant and from industry to industry, but planning is essential and should aim at preventing failure from occurring. However, unnecessarily intrusive maintenance should be avoided. If the equipment is running smoothly and no signs of defects or malfunctions are noticed in the daily routine of cleaning and inspection or through condition monitoring, then it is pointless to strip and rebuild the machine simply because it has been laid down as part of the maintenance plan. Highly sophisticated sensors and software are available to forestall failure in all types of rotating machinery, and they are non-intrusive. However, in TPM the best condition monitor is an operator who is in harmony with his equipment and who has a sense of ownership of that equipment. This leads on to the third principle.

## 4.3   Third principle: establish autonomous maintenance

All three cycles of the TPM improvement plan (see Figures 3.9 and 3.10) involve the principle of autonomous maintenance, or self-determined asset care. In the Japanese TPM approach, there are seven steps of autonomous maintenance (see Chapter 2). These are given below together with the necessary emphasis and differences developed in the Western TPM approach, as reflected in the nine-step TPM improvement plan.

*Initial cleaning*

This starts with the five Ss mentioned in Chapter 2. The cleaning of machines and production plant gives operators an insight, which they never had before, into the condition of their machines. They can therefore use their eyes, ears, nose, mouth and hands to help their maintenance colleagues. By working together as a team they can ensure effective asset care and release maintenance people for tasks requiring a higher level of training and skill. The full implications of the cleaning regime cannot be over-emphasized because ultimately it leads to the reform of the whole production process. To understand this clearly it pays us to look again at the Japanese five Ss and the Westernized CAN-DO approach, and the way in which their application leads to fundamental changes in the workplace.

The Japanese five Ss emphasize the concept of keeping things under control.

### Seiri (organization)

This is the practice of dividing needed and unneeded items at the job site and quickly removing the unneeded. It also means integrating material flow with the best known operational methods.

To better understand the meaning of unneeded items, these can be divided into three different aspects:

- defective products
- not useful items
- not urgent objects, right now.

There are six recommended steps in *seiri* with their own targets for improvement:

- stock, inventory
- tools, jigs
- dies
- containers, pallets
- conveyors, trucks, forklifts
- space.

### Seiton (orderliness)

This means orderly storage, putting things in the right place. Those things can then easily be found, taken out and used again when they are needed. It doesn't simply mean lining things up neatly: it means there is a place for everything, and everything should be in its place! The locations of equipment, tooling and material are clearly defined, displayed and maintained.

### Seiso (cleaning)

This refers to cleaning the workplace regularly, to make work easier and to maintain a safe workplace.

### Seiketsu (cleanliness)

This means being aware of the need for maintaining a clean workplace, not just through cleaning programmes but through ensuring that spillage of liquids and dropping of material, packaging etc. is avoided.

### Shitsuke (discipline)

This is to formalize and practise continuously the above items each day as you work, to have the discipline to always work to these principles.

In WCS International we have developed an eleven-step plant-wide clear and clean exercise for our clients to put the philosophy of the five Ss or CAN-DO into practice. This is often implemented *shortly after initial TPM pilot equipment projects have been launched*, in order to get everyone involved at an early stage. It is not used as a forerunner or start of TPM as is the usual case with the Japanese approach. The Japanese seem quite prepared to spend six to twelve months cleaning up a plant. In the Western world we do not quite have the same level of patience, and we need to experience early live equipment examples called pilots in order to illustrate, prove and believe in the TPM process.

The plant clear and clean process is described as follows.

## Clear out

1  Zone the plant into clear geographical areas with clear management responsibility. (See the plant plan for your shift's responsibility area.)
2  Carry out a first-cut physical run for items that can be immediately thrown away *today* because it is *obvious* they are not needed.
3  Carry out a second red-tag/red-label/red-sticker run which needs to be more structured and thoughtful.
4  It is obvious that if you are to get rid of a great many items, you will need a great many waste disposal containers (say six strategically placed skips). Some items will be wanted but are in the wrong place: 'There must be a place for everything, and everything must be in its right place.'
5  For items to be in the right place we need to paint clear gangways and clear marks on the floor for *anything* mobile (i.e. stillages, raw material, work in progress etc.). Correct racking, shadow boards, labelling and other visual storage aids will form an important part of this stage.
6  Keep the workplace organization under a permanent microscope.

## Clean up

7  Do the obvious sweeping and vacuuming of the work area.
8  Inspect and clean every square centimetre of the equipment. Remember: *every square centimetre*.
9  Identify the points of accelerated deterioration. Where are the leakages and spillages occurring, and *why*? Ask 'why?' five times.
10  Get to the *root causes* of dust, dirt and scattering and *eliminate* those reasons. We will achieve a dust-free plant if – and only if – we achieve this step. All the previous nine steps are useless unless step 10 is achieved.
11  Revisit steps 1 to 10 and continuously improve.

### Counter-measures at the source of problems

Cleaning, checking, oiling, tightening and precision checking of equipment on a daily basis enable operators to detect abnormalities as soon as they appear. From then on operators learn to detect problems and to understand the principles and procedures of equipment improvement. To set this into perspective we can list some examples of situations where the operators have *not* been trained to be equipment conscious:

- dirty or neglected equipment
- disconnected hoses
- missing nuts and bolts producing visible instability
- steam leaks and air leaks
- air filter drains in need of cleaning
- jammed valves
- hydraulic fluid and lubricating oil leaks
- measuring instruments too dirty to read
- abnormal noises in pumps and compressors.

These are glaring examples of a failure to maintain the most basic equipment conditions, but we are deluding ourselves if we believe such situations never arise: they do! Even brand new equipment, if neglected, will rapidly deteriorate (i.e. after just a few days) and its performance and output will drop as a consequence.

### Use of visual management techniques

When the equipment has been cleaned and the weaknesses have been found and corrected, the next phase of the TPM process is to draw attention to the right way of doing things by clear visual aids. This is error proofing: to make it easy to do things right, difficult to do things wrong (Figure 4.6). Some examples of visual marking to induce ease of inspection, discipline, order and tidiness are as follows:

- Where sight glasses are used, make sure that they are clean and that the high and low points are boldly marked and colour coded so that they can be seen at a glance.
- Mark gauges green or red for 'go' or 'no go'.

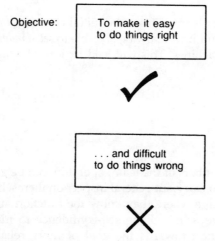

**Figure 4.6** *Improvements*

- Use small windmills to indicate extraction fans and motors working.
- Indicate the correct level on oil bottles as a maximum and a minimum. An elastic band on the bottle will show the level at the previous check, to give the *rate* of use.
- Use line indicators on bolts and nuts to show position relative to their fixture base.
- Provide inspection windows for critical moving parts.
- Colour tag clearly those valves which are open and those which are closed.
- Highlight critical areas which must be kept scrupulously clean.
- Identify covers which are removable by colour coding them.
- Where there is an agreed inspection routine, number in sequence those points which require attention.
- Prepare quality colour photographs of equipment standards and ensure that these are readily accessible to operators.
- Make up shadow boards for tools and spares so that the correct location of every item is immediately apparent.
- Indicate the correct operation of machines by instructions and labels which are visible *on the machine,* kept clean and accessible.
- Display charts and graphs adjacent to the equipment to show standards and to indicate progress towards objectives.

Having completed the first two steps towards autonomous maintenance, operators will have learned to detect problems and to understand the principles and procedures of equipment improvement. They can now take the next steps.

### Cleaning and lubrication standards

Much will have been learned from the initial cleaning, orderliness and discipline procedures and it will now be possible to *set standards* for the ongoing care of plant and machines. This will lead logically towards the next step.

### General inspection

Helped by other members of the team, operators can be guided towards the point where they can carry out general inspection themselves. They will then have reached the stage where they know the function and structure of the equipment and have acquired the self-confidence to make a much more significant contribution towards the goal of more reliable machines and better products.

## Autonomous inspection

As the phrase implies, operators can now carry out self-directed inspection routines which they have defined.

## Organization and tidiness

Initial cleaning and the application of the CAN-DO philosophy will by now have worked through and started to have major effects. In parallel with this, operators will have reached the stage where they can take responsibility for performing autonomous inspection – always within the limits of their skills, experience and training and always backed, where necessary, by their maintenance colleagues. They will have developed an understanding of the relationship between equipment accuracy and product quality. This leads to the final step.

## Full autonomous maintenance

At this stage operators will be equipped to maintain their own equipment. This will include cleaning, checking, lubricating, attending to fixtures and precision checking on a daily basis. They are now equipped to apply their newly developed skills and knowledge to the vital task of continuous improvement.

The key point of emphasis in developing these asset care routines is *empowerment*. The operators' and maintainers' own ideas are encouraged and adopted on the basis that 'If it is my idea and it is embodied in the way in which we operate and look after our equipment, then I will stick with it!' On the other hand, 'If it is imposed from above then I might tick a few check boxes, but I won't actually do anything!' The progress from cleaning to full autonomous maintenance is illustrated in Figure 4.7.

The condition cycle of the TPM improvement plan (Figures 3.9 and 3.10) moves through the following steps:

1 criticality assessment
2 condition appraisal
3 refurbishment programme
4 future asset care.

These cover the seven steps of autonomous maintenance *but* provide the increased structure and discipline which we demand in the Western world to link the process to the measurement and improvement cycles (steps 5 to 9 inclusive) and to *prove* this total nine-step process by setting up TPM pilot projects. This process is fully developed in Chapters 5 and 6. Of course, there must be life after pilots, and this is explained in Chapters 7 and 8.

One of the key points that the reader should now appreciate is that no rocket science is involved in the concepts of TPM: it is basically sound

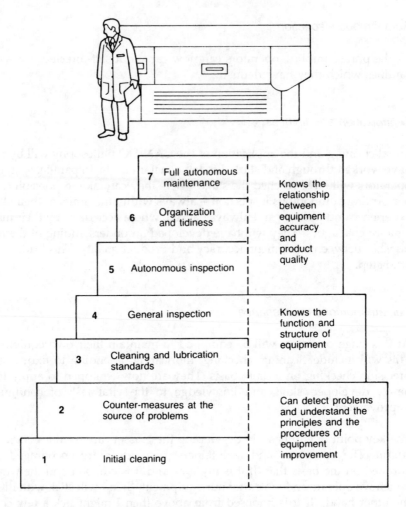

**Figure 4.7**  *Seven steps for developing autonomous maintenance*

common sense. The reality of implementation is, of course, not a straightforward matter, as we shall see in later chapters. Remember, it has taken most manufacturing and process plants ten to twenty years to get to their current state. So, realistically, it is going to take more than a few months to recover the situation.

## 4.4  Fourth principle: training and education

The whole emphasis of Western TPM, and hence the three-cycle, nine-step TPM improvement plan, is geared to taking its participants – whether the chief executive or the operators and maintainers – on a journey from innocence to excellence (Figure 4.8), or if you prefer, from unconscious incompetence to unconscious competence (Figure 4.9)!

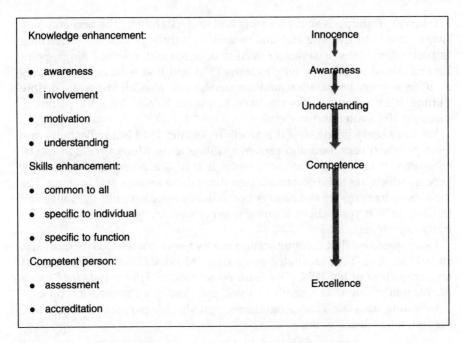

**Figure 4.8** *TPM awareness and training: innocence to excellence*

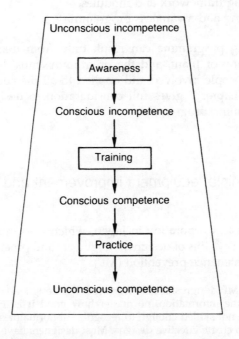

**Figure 4.9** *TPM awareness and training: the path to unconscious competence*

Likewise, training is about learning and understanding. The best and only way to *retain* our learning and understanding is through *experience*. Learning is best retained by a series of well-structured and relevant single-point lessons. There is only one way to learn TPM and that is to go and actually *do it*! However, preparation and awareness are also all-important before putting it into practice. So we have to proceed with thought, planning, care and 100% attention to detail.

We have said already, and it is worth repeating: TPM is a philosophy or a concept which uses hard and proven enabling tools. Moreover, TPM can be measured. It is total quality with teeth! It is also a powerful involvement process which has to be communicated through awareness and training. The benefits of training are not always immediately apparent, but (as suggested in Chapter 3) if you believe training is expensive, try ignorance: that can be *really* expensive!

Remember also that communication is a two-way continuous process, and so is TPM. As a TPM facilitator in Vauxhall Motors, Ellesmere Port, said at the completion of his TPM pilot team presentation: 'TPM stands for "today people matter", and as a result you will get "totally pampered machines".'

In setting up your TPM programme, you should give clear thought and definition to the following:

1   Establish purpose of training.
2   Establish training objectives.
3   Agree method of approach.
4   Set up training framework and modules.
5   Design training and awareness programme.

The total training programme can result only from detailed study of a particular company or plant, and the programme must be tailored to the needs of all the people involved and applied to all the equipment they are working with. Chapter 7 gives full consideration to designing your own tailored TPM training programme.

## 4.5  Fifth principle: equipment improvement and maintenance prevention

The Nachi-Fujikoshi Corporation in Japan, which uses much of the equipment it produces, says this of designers, engineers and procurement staff and the need for maintenance prevention (MP):

> Unless good MP design skills are nurtured, the designers will not understand how to use the information, no matter how good it is. Furthermore, if the designers are not skilled enough to recognize abnormalities as such, they will not be able to create effective designs. Most designers have little work experience in equipment operation and maintenance, so they do not think in terms of autonomous maintenance and maintainability. However, they can overcome these weaknesses and build MP design skills by:

- visiting the factory floor and hearing what the equipment operators and maintenance staff have to say
- studying equipment that has been improved as a result of autonomous maintenance or quality maintenance activities and listening to project result announcements made by TPM circles
- getting hands-on experience in cleaning, lubricating, and inspecting equipment
- conducting several MP analyses based on checklists.

MP designers should have their MP design knowledge and skills evaluated in order to identify remaining weaknesses, facilitate self-improvement, and acquire on-the-job training in more advanced skills.

There is an extraordinarily powerful commercial advantage to a company when this vital pillar and principle of TPM can be mobilized and used to maximum effect. Designers, engineers, technologists, procurement, finance, operations and maintenance will then work as essential partners in the drive to improve the company's overall equipment effectiveness by eliminating many of the reasons for poor maintainability, operability and reliability at source (i.e. at the equipment design, engineering and procurement stage).

The Western TPM principle of core TPM teams supported by key contacts (see Figure 3.3) reflects the importance of MP both in retrofit on existing assets and for the next generation of equipment. Figure 4.1 shows that this vital pillar is also focused through the measurement and improvement cycles of the nine-step TPM improvement plan. Chapter 9 also devotes attention to the subject of TPM for equipment designers, specifiers and planners.

# — 5 —

# *TPM Improvement Plan*

TPM is about maximizing the overall effectiveness of equipment through the people that operate and maintain that equipment. In order to provide the essential link between equipment and people it is essential to identify a clear set of phases and steps which together make up the TPM improvement plan.

As outlined in Chapter 3, there are three phases to the plan:

- the *condition cycle*, which establishes the present condition of the equipment and identifies the areas for improvement and future care
- the *measurement cycle*, which assesses the present effectiveness of the equipment and provides a baseline for the measurement of future improvements
- the *improvement cycle*, which moves equipment effectiveness forward along the road to world class performance.

For convenience, Figures 5.1 and 5.2 repeat Figures 3.9 and 3.10. They show the interrelationships between the three cycles and the nine steps outlined in Chapter 3 and described in more detail in this chapter.

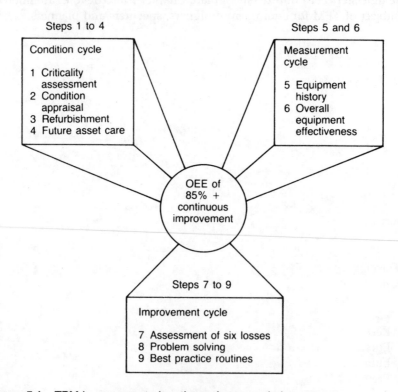

**Figure 5.1** *TPM improvement plan: three phases and nine steps*

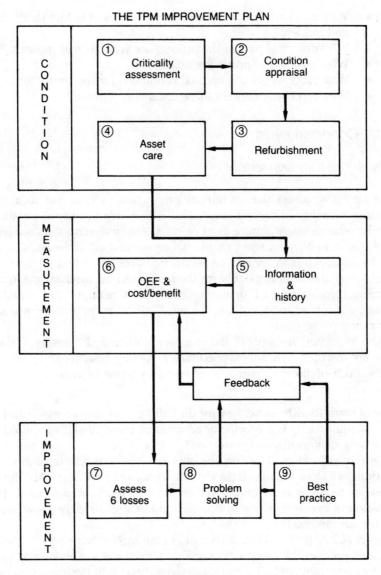

**Figure 5.2** *TPM improvement plan sequence*

Throughout the TPM improvement plan five themes prevail, as discussed in Chapter 3:

1 Restore before improve.
2 Pursue ideal conditions.
3 Eliminate minor defects.
4 Prevent causes of dust, dirt and scattering.
5 Ask why fives times:

- Why don't we know the true consequences of failure (both obvious and hidden)?
- Why does this part of the process not work as it is meant to?
- Why can't we improve the reliability?
- Why can't we set the optimal conditions for the process?
- Why can't we maintain those optimal conditions?

## 5.1   Condition cycle

### Step 1: Criticality assessment

The aim is to assess the equipment production process and to agree the relative criticality of each element. This will enable priority to be allocated for the refurbishment, future asset care and improvement of those elements most likely to have an effect on overall equipment effectiveness.

The approach is to review the production process so that all members of the team understand (probably for the first time!) the mechanisms, controls, material processing and operating methods. Operators and maintainers must be involved in identifying the most critical parts of the process from their own perspective.

The important elements of the process are identified: some typical examples are electrics, hydraulics, pneumatics, cooling systems and control systems. Each of these elements is assessed in terms of criteria such as the following:

*Ease of repair* (EOR)   How easy (or difficult) is it to gain access to and work on the equipment? Is a specific or an unusual blend of skills required to do necessary diagnostics and correction?

*Reliability* (R)   How do those directly involved rate the reliability of the equipment? Does it suffer from chronic/continual problems? Are the problems sporadic or recurring? What is the frequency of problems? Is it a question of breakdowns or of not quite operating correctly (slow running, minor adjustments)?

*Product quality* (PQ)   What impact does malfunction have on the quality of finished products? Is the product quality impact quickly and easily identifiable, or apparent only after further deterioration? Will the impact on quality not become noticeable until later in the process?

*Throughput velocity* (TPV)   what impact does the equipment condition and performance have on the throughput velocity of the product?

*Loss of production* (LOP)   What is the effect on the performance of the total system, production line or department if the equipment is not operating correctly, is broken down, is unreliable, or is difficult to maintain?

*Safety* (S)   What is the likely impact on safety considerations if the equipment is not quite right or is proving to be unreliable?

*Environment* (ENV)   What impact does the condition of equipment have on the environment, both locally and on a wider front?

*Cost* (C)   What are the financial consequences of failure, or unreliability, or poor performance? Both obvious and hidden costs should be considered.

The significance of each of the criteria is assessed and allocated a score according to impact on the process elements selected (electrics, hydraulics etc.). For example, for PQ, TPV, LOP, S, ENV and C the score is 1 if the impact is low and 3 if the impact is high; for EOR the score is 1 for easy and 3 for difficult; and for R the score is 1 for high and 3 for low.

A typical matrix form for recording process elements and criteria scores is shown in Figure 5.3. The right-hand (totals) column enables priority to be applied to those elements most affected. This is further illustrated in Figures 5.4 and 5.5.

| EQUIPMENT DESCRIPTION | 1-3 RANKING AS IMPACT ON: | | | | | | | | |
|---|---|---|---|---|---|---|---|---|---|
| | M | R | PQ | ENV | TPV | KOE | S | C | TOT |
| | | | | | | | | | |

| where: | | | | 1 | 3 |
|---|---|---|---|---|---|
| M | = | Maintainability | | Easy | Difficult |
| R | = | Reliability | | High | Low |
| PQ | = | Product Quality | | Low | High |
| ENV | = | Environment | | Low | High |
| TPV | = | Throughput Velocity | | Low | High |
| KOE | = | Knock on Effect | | Low | High |
| S | = | Safety | | Low | High |
| C | = | Cost | | Low | High |

**Figure 5.3** *Criticality assessment matrix form*

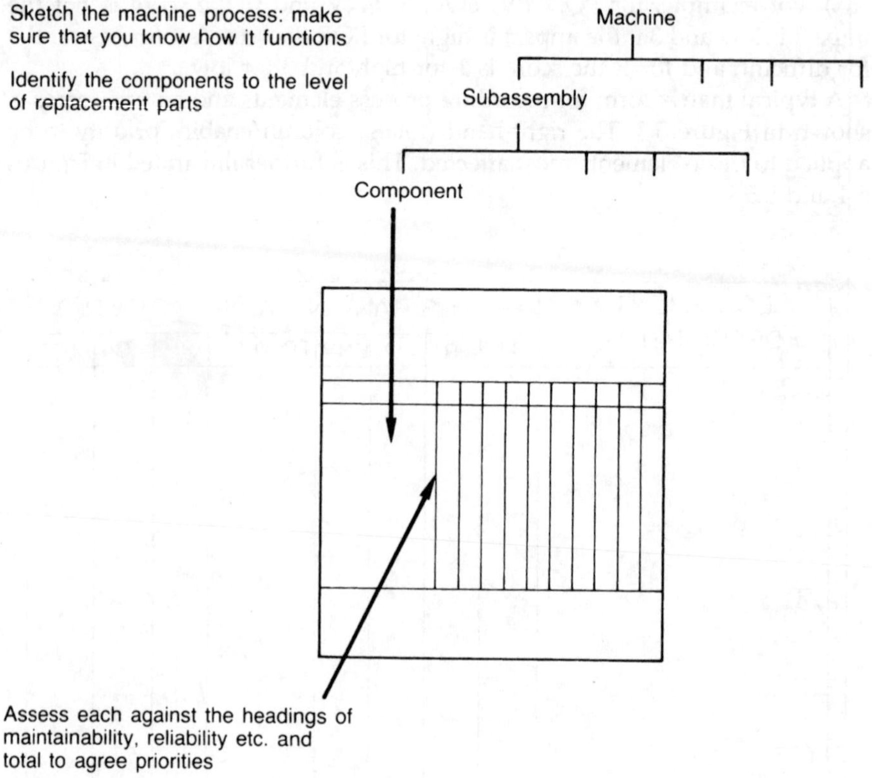

Sketch the machine process: make
sure that you know how it functions

Identify the components to the level
of replacement parts

Machine

Subassembly

Component

Assess each against the headings of
maintainability, reliability etc. and
total to agree priorities

**Figure 5.4**  *Stages in criticality assessment*

The main outputs from the criticality assessment process are that it:

- starts the teamwork building between operators and maintainers;
- results in a fuller understanding of their equipment;
- provides a checklist for the condition appraisal;
- provides a focus for the future asset care;
- highlights weaknesses regarding: operability; reliability; maintainability.

*Step 2: Condition appraisal*

The objective is to make use of the same criticality assessment elements and components in order to assess the condition of equipment and to identify the refurbishment programme necessary to restore the equipment to maximum effectiveness.

| Equipment Description | | | | | | | | | | PRIORITY |
|---|---|---|---|---|---|---|---|---|---|---|
| *RH DOOR HINGE REINFORCEMENT* 1 - 3 ranking as impact on : *WELDER* | M | R | PQ | CO | TPV | KOE | S | C | TOT | |
| A-C MODICON 894145 PLC | 1 | 1 | 1 | 3 | 3 | 3 | 1 | 3 | 16 | (1) |
| I/O MODULES DEP 2160 | 1 | 1 | 1 | 1 | 3 | 3 | 1 | 3 | 14 | (2) |
| I/O MODULES DEP 216 | 1 | 1 | 1 | 1 | 3 | 3 | 1 | 3 | 14 | (2) |
| DOOR SW RIT. 52 2586 | 1 | 1 | 1 | 1 | 1 | 1 | 1 | 1 | 8 | (4) |
| FLOURESC. TUBE F 20W - 0M/RS | 1 | 1 | 1 | 1 | 1 | 1 | 1 | 1 | 8 | (4) |
| COOLING FANS RS 509-068 | 3 | 1 | 1 | 1 | 1 | 1 | 1 | 1 | 10 | (3) |
| FILTERS | 1 | 1 | 1 | 1 | 1 | 1 | 1 | 1 | 8 | (4) |
| PUSH BUTTON C. Z - BU06 | 1 | 1 | 1 | 1 | 1 | 1 | 1 | 1 | 8 | (4) |
| Z - BZ-B 102WK | 1 | 1 | 1 | 1 | 1 | 1 | 1 | 1 | 8 | (4) |
| Z - BZ-B 101NK | 1 | 1 | 1 | 1 | 1 | 1 | 1 | 1 | 8 | (4) |
| INDICATOR LAMP Z - BU6 | 1 | 1 | 1 | 1 | 1 | 1 | 1 | 1 | 8 | (4) |
| P/BUTTON SW Z - BU06 | 1 | 1 | 1 | 1 | 1 | 1 | 1 | 1 | 8 | (4) |
| E/STOP BUTTON | 1 | 1 | 1 | 1 | 3 | 3 | 3 | 1 | 14 | (2) |
| | | | | | | | | | | |
| | | | | | | | | | | |
| | | | | | | | | | | |

| Rating : | | | - 1 - if: | - 3 - if: |
|---|---|---|---|---|
| M | = | Maintainability | Easy | Difficult |
| R | = | Reliability | High | Low |
| PQ | = | Product Quality | Low | High |
| CO | = | Change Over | Low | High |
| TPV | = | Throughput Velocity | Low | High |
| KOE | = | Knock-On Effect | Slow | Rapid |
| S | = | Safety | Low | High |
| C | = | Cost | Low | High |

**Figure 5.5** *Completed criticality assessment matrix*

Each heading will have been subdivided as necessary: for example, the electrical section may contain power supply, control panel, motors and lighting.

Under each of the subdivisions of the equipment being studied, four categories should be established:

- satisfactory
- broken down
- needs attention now
- needs attention later.

An example of the outcome of a condition appraisal study is shown in Figure 5.6.

### Step 3: Refurbishment

The objective of the refurbishment programme is to set up a repair and replacement plan, based on the condition appraisal, and indicating the resources needed. Getting the equipment back to an acceptable level is a prerequisite to the pursuit of ideal conditions.

The plan will provide a detailed summary of actions to be coordinated by the team, which will include:

- dates and timescales
- resources (labour, materials, time)
- cost estimates
- responsibilities
- control and feedback (management of change).

A typical summary table of refurbishment costs and manhours required for a group of critical machines is shown in Figure 5.7

The chart in Figure 5.8 gives details of action required on a specific item of equipment. It allocates responsibility for the various tasks and nominates individuals to carry out the work; it also embodies a simple visual indication of progress with the work.

The refurbishment programme is concerned not just with clearly identifiable repair work but with the many small weaknesses identified by the cleaning and CAN-DO approach, such as missing bolts, leaks, temporary repairs and over/under-lubrication, and it highlights critical points for regular attention.

CONDITION APPRAISAL - TOP SHEET

MACHINE N° : YM 56694
DATE INSTALLED : 20.07.91
COMMISSIONED : 01.08.91
WARRENTY ENDS : 01.08.92
LOCATION CODE : K19
PLANT PRIORITY : HIGH
GENERIC GROUP : RH. FLOOR Assy
P.O. NUMBER : 0619862
COMMON EQUIPMENT:

DESCRIPTION: RH. FRONT DOOR
HINGE RE-INFORCEMENT
CO₂ WELDER
MARKER: ESTIL
MANUFACTURE SERIAL N°: 0766/01/00
MARKERS N°:
EQUIPMENT STATUS: OPERATIONAL
EQUIPMENT AVAILIBILITY: NIGHT & DAY

GENERAL STATEMENT OF RELIABILITY

GENERAL STATEMENT (
THE ACCESS TO THE R/A;
BECAUSE OF THE SMOG h
ALSO THE WELDING RECTI

CONDITION APPRAISAL   1 of....

MACHINE DESCRIPTION:

| ASSET N°: | YEAR OF PURCHASE: | APPRAISAL BY: |
| MACHINE N°: | LOCATION: | APPRAISAL DATE: |

| ITEM N° | APPRAISAL RATING BY SUB ASSET | NOT APPLICABLE | SATISFACTORY | BROKEN DOWN | NEEDS ATTENTION NOW | NEEDS ATTENTION LATER |
|---|---|---|---|---|---|---|
| | | "X" AS REQUIRED | | | | |
| 1 | ELECTRICAL | | | | | |
| | A – POWER SUPPLY TO MACHINE | | X | | | |
| | B – PANEL | | | | X | |
| | C – CONTROL/LS | | | | X | |
| | D – CONTOL CIRCUITS | | | | X | |
| | E – MOTORS | | | | X | |
| | F – MACHINE LIGHTING | X | | | | |
| | G – TRUNKING | | | | X | |
| 2 | MECHANICAL | | | | | |
| | A – SPINDLE HOUSINGS/GEARBOXES | | | | X | |
| | – SEALS | | | | | X |
| | – BEARINGS | | | | | X |
| | – GEARS | | X | | | |
| | B – SLIDEWAYS/TABLES | | | | X | |
| | – WORKPIECE | | | | X | |
| | – TOOLHOLDER | | | | X | |
| | C – SCREWS/RAMS/SLINED SHAFTS | X | | | | |
| | D – PNEUMATICS | | | | X | |

**Figure 5.6**  *Example of condition appraisal study*

| Asset No. | Machine Description | SUMMARY OF TOP 20 CRITICAL MACHINES - Refurbishment Program - | | | | | |
|---|---|---|---|---|---|---|---|
| | | Estimated refurb't cost | | | Manhours work | | |
| | | Req'd now | Req'd later | Total | Req'd now | Req'd later | Total |
| 873 | RICHMOND BED DRILL | 70 | 800 | 870 | 39 | 160 | 199 |
| 929 | DEVLIEG 5 PALLET | 16135 | 6250 | 22385 | 41 | 98 | 139 |
| 134 | MAXICUT | 10 | 800 | 810 | 9 | — | 9 |
| 871 | SNOWGRINDER 65/7·5 | 250 | 10600 | 10850 | 14 | 10 | 24 |
| 876 | N°2 MILL | 555 | 16100 | 16655 | 58 | 89 | 147 |
| 877 | RADYNE BED HARDENER | 1260 | 990 | 2250 | 40 | 92 | 132 |
| 443 | VOUMARD | 30 | 80 | 110 | 5 | 8 | 13 |
| 477 | WORKMASTER | 2740 | 3850 | 6590 | 78 | 40 | 118 |
| 461 | NAXOS | 245 | 35 | 280 | 10 | 14.5 | 24.5 |
| 652 | LANDIS | 370 | — | 370 | 24 | 32 | 56 |
| 847 | DEVLIEG | 190 | 2760 | 2950 | 24 | 122 | 146 |
| 849 | DEVLIEG | 60 | 17760 | 17820 | 17.5 | 212 | 229.5 |
| 873 | N°4 MILL | 85 | — | 85 | 11.17 | 40 | 51.17 |
| 958 | HYDRO 540 | 160 | 8270 | 8430 | 21.67 | 4 | 25.67 |
| 925 | CNC 650 G·AXIS I | 3330 | 230 | 3560 | 29 | 40.17 | 69.17 |
| 926 | CNC 650 C-AXIS II | 190 | 790 | 980 | 17.28 | 40.17 | 57.42 |
| 941 | CNC 650 | 150 | 100 | 250 | 6 | 4.17 | 10.17 |
| 541 | SNOW GRINDER 170/20 | 235 | 10550 | 10785 | 25 | 14 | 39 |
| 649 | SNOW GRINDER 170/20 | 195 | 12520 | 12715 | 46.5 | 51 | 97.5 |
| 875 | N°1 MILL | 365 | 14000 | 14365 | 13 | 12 | 25 |
| | TOTAL : | 22,625 | 106,485 | 133,110 | 529.1 | 1,083 | 1,161.1 |

**Figure 5.7** *Refurbishment example for a group of machines: how costs can be spread*

| Unit EA SUBS | Team LH FRONT DOOR A.B.C | | *T.P.M.* **Refurbishment Chart** | | | Manager G.BOOTH R.O'NEIL | Supervisor F.ROBINSON B.SEYMOUR | Date 30.11.92 |
|---|---|---|---|---|---|---|---|---|
| Division 631 | Operation CO2 Welder | | | | | | | |
| No. | Item to be Refurbished | Action Required | Responsibility | Completion Date | Champion | Check Date | | Progress |
| 1 | DOUBLE ACTUATING CYLINDER AIR FEED | REPAIR / REPLACE | D.HUDSON | 6.12.92 | K.DOYLE | | | ◐ |
| 2 | HOSE CARRIER CHAFFING | RE-LOCATE | R.EDWARDS | 6.12.92 | K.DOYLE | | | ◐ |
| 3 | SLIDE MOUNTING BOLTS | REPLACE / CHECK TIGHTNESS | C.TAYLOR | 6.12.92 | I.HOWIE | | | ◐ |
| 4 | FLEXIBLE GAS SUPPLY PIPE | CHANGE FROM BLUE TO ORANGE | R.PRICE | 6.12.92 | R.GEBBINGTON | | | ◐ |
| 5 | CONTROL PANEL COOLER FAN | RE-SITE | L.PHILLIPS | 6.12.92 | H.COOPER | | | ◐ |
| 6 | CONTROL PANEL LIGHTS | REPLACE | H.COOPER | 6.12.92 | T.FOLEY | | | ◐ |
| 7 | ISOLATOR | IDENTIFY | H.CRAIG | 6.12.92 | I.LAMEN | | | ◐ |
| 8 | SOLENOID VALVES | IDENTIFY | I.LAMEN | 6.12.92 | R.CASSIDY | | | ◐ |
| 9 | RECTIFIER CABLES | TIDY / RE-ROUTE | C.WEATHERLEY | 6.12.92 | H.COOPER | | | ◐ |
| 10 | WIRE FEED UNIT | REPAIR | L.PHILLIPS | 6.12.92 | R.PRICE | | | ◐ |
| 11 | ANTI-SPATTER RESERVOIR | INVESTIGATE WHY INOPERATIVE / REPAIR | R.PRICE | 6.12.92 | D.CARROLL | | | ◐ |
| 12 | GENERAL MACHINE | CLEAN / PAINT | I.HOWIE | 6.12.92 | T.FOLEY | | | ◐ |

Fill in first segment when a refurbishment has been flagged | Fill in second segment when action is defined and responsibility given | Fill in third segment when refurbishment has been completed | Fill in forth segment when refurbishment has been tested

CO2 Mig Welding M/C :
Labour costs :

| | |
|---|---|
| 2 x 16 hours = 32 hours at £ 7.50 = | £ 240.00 |
| 1 x 13 hours = 13 hours at £ 6.50 = | £  84.50 |
| **Total Labour =** | **£ 324.50** |

Parts :

| | |
|---|---|
| New seals to clamp cylinder = | £  15.00 |
| 6 PX leads = | £  60.00 |
| Water flow gauge = | £  10.00 |
| New air ducting = | No cost |
| Water pressure gauge  = | £ 113.00 |
| Total parts = | £ 198.00 |
| **Total Mig Welding M/C =** | **£ 522.50** |

**Figure 5.8**   *Refurbishment example for a specific item*

*Step 4: Asset care*

Once refurbishment of an item of equipment has been carried out, a future asset care programme must be planned to ensure that the machine condition is maintained. It is therefore necessary to establish:

- cleaning and inspection routines
- checking and condition monitoring methods and routines
- planned, preventive maintenance and service schedules.

For each of these we must develop:

- improvements to make each task easier
- visual techniques to make each task obvious
- training to achieve consistency between shifts.

It is important to distinguish between natural and accelerated deterioration. In the course of normal usage. Natural deterioration will take place even though the machine is used properly. Accelerated deterioration arises from outside influences. These are equipment based, i.e. failure to tackle the root causes of dust, dirt and contamination; and operator based, i.e. failure to maintain basic conditions such as cleaning, lubricating and bolting, and also human operational errors.

Figures 5.9 and 5.10 show how the care of assets may be broken down into elements which reflect the first three steps of the condition cycle: criticality,

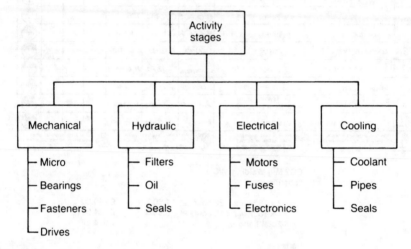

**Figure 5.9**   *Stages in asset care*

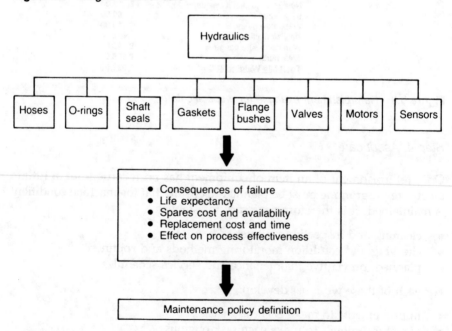

**Figure 5.10**   *Breakdown of asset care for hydraulics maintenance*

condition and refurbishment. Figure 5.11 illustrates the interrelationship between operational and technical aspects of asset care. Some key points for consideration in asset care are shown in Figure 5.12.

The question of training is developed fully in Chapter 7, but some key approaches are illustrated in Figure 5.13. A training schedule form is shown in Figure 5.14. This schedule is completed through a series of single-point, on the job lessons.

A practical example of daily cleaning and inspection is given in Figure 5.15. This shows the checks to be made in a MIG welding cell for each shift during the working week, and records all the daily checks made by the operators. A material usage chart for the same production process is shown in Figure 5.16. The key point is that the operators and maintainers have developed these asset care routines on the basis that 'If it's my idea, I will stick with it.'

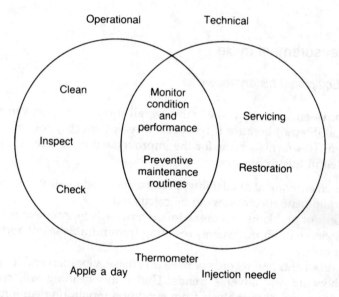

**Figure 5.11**  *Relationship between operational and technical aspects of asset care (see also Figure 3.2)*

- Cleaning and inspection
- Checks and monitoring
- Preventive maintenance and servicing

For each task make it:

- *Easy* by simple improvements
- *Obvious* using visual techniques
- *Consistent* by effective training

**Figure 5.12**  *Key points in asset care*

| Technique | Learning points | Improvements | Training |
|-----------|-----------------|--------------|----------|
| • Cleaning | Accelerated deterioration | Highlight vulnerability | |
| | Cleaning is inspection | Make it easy | Video |
| • Inspection | What is the effect? | Provide tools/equipment | |
| • Checking | Check condition | Establish standards | |
| | Check performance | Establish parameters | |
| | | | Instruction |
| • Condition monitoring | What are the signs? | Make change obvious | |
| | Using our senses | Make detection easy | |
| | Using instruments | Provide tools/equipment | |
| • Planned preventive maintenance | What is to be done? | Make it accessible | Single-point lesson |
| | How do we do it? | Make it maintainable | |
| | Who and when? | Clear responsibility | |
| • Servicing | How do we manage it? | Clear instructions | |

**Figure 5.13**   *Role of training in asset care*

## 5.2   Measurement cycle

*Step 5: Equipment history record*

This is the essential prerequisite to the overall equipment effectiveness (OEE) calculation (Step 6) because it records the recent effectiveness of an equipment item. This forms a basis for the improvement cycle (see later).

Included in this record are:

- Data on equipment availability, performance and quality to enable overall equipment effectiveness to be calculated.
- Records of problems and breakdowns as a basis for problem solving and as evidence of improvements resulting from refurbishment and ongoing asset care.
- Measurements and records of pressure, noise, vibration and temperature to show up any adverse trends. Under this heading will come data collection and data analysis from condition monitoring equipment.

All of this information will have a direct bearing on the ongoing asset care and improvement programme.

A typical equipment history record is shown in Figure 5.17 and records of this type will form the basis of the OEE calculation described below.

*Step 6: Overall equipment effectiveness*

The OEE formula is at the heart of the TPM process. It is soundly based on measurable quantities and enables progress to be quantified as the organization embraces TPM with all its implications. The formula enables calculation of two parameters:

# TOTAL PRODUCTIVE MAINTENANCE PROGRAM

## R/H FRONT DOORLINE MIG WELDING CELL

TEAM LEADER.......... T FOLEY          OPERATOR TRAINING SCHEDULE

| TEAM MEMBERS NAME | CHECK AIR GAUGE | CLEAN M/C | CHECK Co2 WIRE | CHANGE Co2 WIRE | CLAMP CHECK | AIR. WATER LEAKS | TORCH CHECK | TOP UP FLUID | CLEAN SHROUD | CHECK SMOG HOG | RECORD +RESET CYCLE COUNTER | CHECK PART LEVELS | PARTS ON BOARD | SCRAP LEVEL | REWORK LEVEL |
|---|---|---|---|---|---|---|---|---|---|---|---|---|---|---|---|
| A SEFTON | | | | | | | | | | | | | | | |
| B DEAN | | | | | | | | | | | | | | | |
| P LISLE | | | | | | | | | | | | | | | |
| D CRAVEN | | | | | | | | | | | | | | | |
| I MORRIS | | | | | | | | | | | | | | | |
| B GALLAGHER | | | | | | | | | | | | | | | |
| G SEDDON | | | | | | | | | | | | | | | |
| M LAWTON | | | | | | | | | | | | | | | |
| N MELLING | | | | | | | | | | | | | | | |
| R CROSS | | | | | | | | | | | | | | | |
| L CURRIE | | | | | | | | | | | | | | | |
| D WELLS | | | | | | | | | | | | | | | |
| J HARWOOD | | | | | | | | | | | | | | | |
| S MORRIS | | | | | | | | | | | | | | | |

TRAINED IN PROCEDURES BY MAINTENANCE    CARRIED OUT PROCESS    COMPETENT IN PROCESS    ABLE TO TRAIN OTHERS

**Figure 5.14** *Training schedule form*

# TOTAL PRODUCTIVE MAINTENANCE PROGRAMME
## R/H FRONT DOORLINE CO2 MIG WELDING CELL.
### DAILY CLEANING AND INSPECTION CHECKS

DATE: 28.9.92.

NOTIFY MAINTENANCE

WORK TO BE CARRIED OUT

✓ IF OK
X IF NOT OK

| | SHIFT | CHANGE CO2 WIRE | CHECK PRESSURE SETTING | CLEAN TABLE + TOOLING | CHECK CO2 WIRE REEL + SPARE | CHECK CLAMP HEAD SECURITY | CHECK FOR AIR + WATER LEAKS | CHECK TORCH + HARNESS SECURITY | TOP UP ANTI-SPATTER FLUID | REMOVE SHROUD AND CLEAN | CHECK SMOG HOG LIGHT IS ON | RECORD + RESET CYCLE COUNTER | CHECK MATERIAL LEVELS | CHECK PARTS ON SHADOW BOARD | RECORD SCRAP LEVEL | RECORD REWORK LEVEL | SIGNATURE |
|---|---|---|---|---|---|---|---|---|---|---|---|---|---|---|---|---|---|
| | | X | X | X | | X | X | X | X | | X | X | | X | X | X | |
| MONDAY | D/S | ✓ | ✓ | ✓ | ✓ | ✓ | ✓ | ✓ | ✓ | X | ✓ | 234 | ✓ | X | ✓ | ✓ | Tw |
| | N/S | ✓ | ✓ | ✓ | ✓ | X | ✓ | ✓ | ✓ | X | ✓ | 380 | ✓ | ✓ | 0 | 0 | S.C. |
| TUESDAY | D/S | ✓ | ✓ | ✓ | ✓ | ✓ | ✓ | ✓ | ✓ | ✓ | ✓ | 328 | ✓ | X | 0 | 0 | |
| | N/S | ✓ | ✓ | ✓ | ✓ | ✓ | ✓ | ✓ | ✓ | ✓ | ✓ | 200 | ✓ | ✓ | 0 | 0 | |
| WEDNESDAY | D/S | ✓ | ✓ | ✓ | ✓ | ✓ | ✓ | ✓ | ✓ | ✓ | | 326 | ✓ | X | 0 | 0 | |
| | N/S | ✓ | ✓ | ✓ | ✓ | ✓ | ✓ | ✓ | ✓ | ✓ | | 345 | X | X | 0 | 0 | S.C. |
| THURSDAY | D/S | ✓ | ✓ | ✓ | ✓ | ✓ | ✓ | ✓ | ✓ | ✓ | ✓ | 103 | ✓ | ✓ | 0 | 0 | |
| | N/S | ✓ | ✓ | ✓ | ✓ | ✓ | ✓ | ✓ | ✓ | ✓ | ✓ | 326 | X | ✓ | 0 | 0 | |
| FRIDAY | D/S | ✓ | ✓ | ✓ | ✓ | ✓ | ✓ | ✓ | ✓ | ✓ | ✓ | L87 | ✓ | ✓ | 0 | 0 | |
| | N/S | | | | | | | | | | | | | | | | |
| SATURDAY | D/S | | | | | | | | | | | | | | | | |
| | N/S | | | | | | | | | | | | | | | | |

**Figure 5.15** Example of daily cleaning and inspection checks

## TOTAL PRODUCTIVE MAINTENANCE PROGRAMME

### R/H FRONT DOORLINE CO2 MIG WELDING CELL

### MATERIAL USAGE CHART

| MATERIAL | WEEK NO. 34 | WEEK NO. 35 | WEEK NO. 36 | WEEK NO. 37 | WEEK NO. 38 | WEEK NO. 39 | WEEK NO. 40 | WEEK NO. 41 | WEEK NO. 42 | WEEK NO. 43 | WEEK NO. 44 | WEEK NO. 45 | WEEK NO. 46 | WEEK NO. 47 |
|---|---|---|---|---|---|---|---|---|---|---|---|---|---|---|
| WELDING WIRE | | | | | | | | | | | | | | |
| WELDING TIP | | | | | | | | | | | | | | |
| SHROUD | | | | | | | | | | | | | | |
| ANTI–SPATTER SPRAY | | | | | | | | | | | | | | |
| DISTILLED WATER | | | | | | | | | | | | | | |
| | | | | | | | | | | | | | | |
| | | | | | | | | | | | | | | |
| | | | | | | | | | | | | | | |

**Figure 5.16** *Material usage chart for example in Figure 5.15*

| Equipment: | CO$_2$ MIG welding cell | | | | VIII Ideal cycle time: 0.5 min/piece | | |
| --- | --- | --- | --- | --- | --- | --- | --- |
| | I | II | III | IV | V | VI | VII |
| Date/shift | Total working time (min) | Planned downtime (min) | Total available time (min) | Actual downtime (stoppages) (min) | Output (no.) | Defects (rework) (no.) | Defects (scrap) (no.) |
| 28.9.92 D/S | 240 | 20 | 220* | 5 | 243 | 0* | 0* |
| 28.9.92 N/S | 240 | 20 | 220 | 0* | 380 | 0 | 0 |
| 29.9.92 D/S | 240 | 20 | 220 | 10 | 328 | 0 | 0 |
| 29.9.92 N/S | 240 | 20 | 220 | 5 | 200 | 0 | 0 |
| 30.9.92 D/S | 240 | 20 | 220 | 0 | 326 | 0 | 0 |
| 30.9.92 N/S | 240 | 20 | 220 | 10 | 345 | 0 | 0 |
| 1.10.92 D/S | 240 | 20 | 220 | 10 | 103 | 0 | 0 |
| 1.10.92 N/S | 240 | 20 | 220 | 5 | 386* | 0 | 0 |
| 2.10.92 D/S | 240 | 20 | 220 | 5 | 187 | 0 | 0 |
| Total | 2160 | 180 | 1980 | 50 | 2498 | 0 | 0 |

* Best scores for use in calculation of best of best.

**Figure 5.17** *Equipment history record for efficiency calculation*

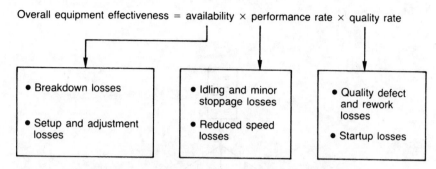

**Figure 5.18**  *Measuring equipment effectiveness*

*Actual effectiveness* of the equipment taking into consideration its availability, its performance rate when running and the quality rate of the product produced. All of these are measured over a period.

*Potential improvement*  The first improvement objective is to obtain constantly, through standardization and stabilization, the best of the best in each of three categories: availability, performance and quality. Beyond this point there must be continuous improvement towards world class levels.

The OEE formula is as follows:

$$\begin{matrix} \text{overall equipment} \\ \text{effectiveness} \end{matrix} = \begin{matrix} \text{availability} \\ \text{of the asset} \end{matrix} \times \begin{matrix} \text{performance rate} \\ \text{when running} \end{matrix} \times \begin{matrix} \text{quality rate of} \\ \text{product produced} \end{matrix}$$

The three factors in the OEE calculation are all affected to various degrees by the six big losses which were outlined in Chapter 3. It is only by a single-minded and sustained attack on these losses that the TPM process can become effective – a change that will be demonstrated by improvements in the OEE.

*Availability* will be affected by breakdown losses and by setup and adjustment losses. (A breakdown requires the presence of a maintenance engineer to correct it.)

*Performance* will be affected by idling and minor stoppages losses and by reduced speed losses. (A minor stoppage can be corrected by the operator and is usually of less than 10 minutes duration.)

*Quality* will be affected by quality defect and rework losses and by startup losses.

In detail the calculation of the three factors whose product determines the OEE is as follows:

$$\text{availability} = \frac{\text{total available time} - \text{actual downtime}}{\text{total available time}} \times 100\%$$

Total available time is total working time less plant downtime allowances (i.e. tea breaks, lunch breaks, personal relief) and planned downtime. Unplanned downtime is caused by breakdowns and changeovers.

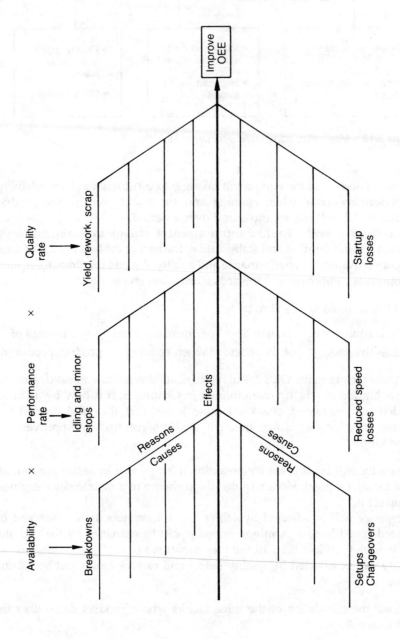

**Figure 5.19** Factors in overall equipment effectiveness

performance = operating speed rate × operating rate

$$= \frac{\text{ideal cycle time}}{\text{actual cycle time}} \times \frac{\text{actual cycle time} \times \text{output}}{\text{operating time}} \times 100\%$$

Ideal cycle time is the cycle time the machine was designed to achieve at 100%. Output is output *including defects*. Operating time is total available time minus unplanned stoppages (i.e. available time).

$$\text{quality rate} = \frac{\text{total output} - \text{number of defects}}{\text{total output}} \times 100\%$$

Measuring OEE was shown diagramatically in Figures 3.5 and 4.4, repeated for convenience as Figures 5.18 and 5.19.

### OEE calculation for welding cell

Calculation of OEE can best be demonstrated by using the values in Figure 5.17. The roman numerals refer to the columns in the figure.

### Average OEE calculation

$$\text{Availability} = \frac{\text{III} - \text{IV}}{\text{III}} = \frac{1980 - 50}{1980} \times 100 = 97.5\%$$

$$\text{Performance rate} = \frac{\text{V} \times \text{VIII}}{\text{III} - \text{IV}} = \frac{2498 \times 0.5}{1980 - 50} \times 100 = 64.7\%$$

$$\text{Quality} = \frac{\text{V} - \text{VI} - \text{VII}}{\text{V}} = \frac{2498 - 0 - 0}{2498} \times 100 = 100\%$$

$$\text{Average OEE} = 0.975 \times 0.647 \times 0.100 \times 100 = 63.1\%$$

### Best of best (target) OEE calculation

The best of best calculation uses the best scores in the period from each column. This gives us a theoretical achievable performance if all of these best scores were constantly achieved. It is our first target for improvement.

$$\text{Availability} = \frac{\text{III} - \text{IV}}{\text{III}} = \frac{220 - 0 \times 100}{220} = 100\%$$

$$\text{Performance rate} = \frac{\text{V} \times \text{VIII}}{\text{III} - \text{IV}} = \frac{386 \times 0.5}{220 - 0} \times 100 = 87.7\%$$

$$\text{Quality} = \frac{\text{V} - \text{VI} - \text{VII}}{\text{V}} = \frac{386 - 0 - 0}{386} \times 100 = 100\%$$

Best of best OEE = $0.100 \times 0.877 \times 0.100 \times 100 = 87.7\%$

*Question* What is stopping us achieving the best of best consistently?
*Answer* We are not in control of the six big losses!

The best of best calculation generates a high confidence level, as each value used of the three elements (availability, performance rate, quality) was achieved at least once during the measurement period. Therefore, if control of the six big losses can be achieved, our OEE will be at least the best of best level.

We can now start putting a value to achieving the best of best performance.

**TPM potential savings for achieving best of best**

| | |
|---|---|
| Cycle time | $A = 30\,\text{s}$ |
| Number of men | $B = 2$ |
| Allowance in standard hours | |
| (personal relief, technical allowance etc.) | $C = 11\%$ |

Credit hours generated per piece

$$X = \frac{(A \times B) + C}{3600\,\text{s}} = \frac{(30 \times 2) + 11\%}{3600\,\text{s}} = 0.0185$$

| | |
|---|---|
| Variable cost per credit hour | $Y = £27.60$ |
| Direct labour cost per piece | $X \times Y = £0.5106$ |
| Current OEE | $D = 63.1\%$ |
| Number of pieces produced | $E = 2498$ |
| Best of best OEE | $F = 87.7\%$ |

Number of pieces produced at OEE 87.7%    $G = \dfrac{F}{D} \times E = 3472$

| | |
|---|---|
| Difference in pieces produced | $G - E = 974$ |
| Potential weekly savings | $= £0.5106 \times 974 = £497$ |
| Potential annual savings (45 working weeks) | $= £22\,365$ |

An alternative to increasing the output potential of 974 pieces per week at best of best is to achieve the same output of 2498 pieces in less time:

Loading time (total available time) was 1980 minutes (33 hours) to produce 2498 pieces at OEE of 63.1%

Loading time to produce 2498 pieces at best of best OEE of 87.7% would be

$$\frac{63.1}{87.7} \times 33 = 23.74 \text{ hours} = 1425 \text{ minutes}$$

Time saving $= 1980 - 1425 = 555$ minutes $= 9.25$ hours

*Simple OEE calculation*

If the foregoing 'live' example seemed a little complicated, let us take the following very simple example.

**Data**

- Loading time = 100 hours, unplanned downtime = 10 hours.
- During run time, output planned to be 1000 units. We actually processed 900 units.

- Of these 900 units processed, only 800 were good or right first time.
- What is our OEE score?

## Interpretation

Availability: actual 90 hours out of expected 100 hours.
Performance: actual 900 units out of expected 1000 units in the 90 hours.
Quality: actual 800 units out of expected 900 units.

## Calculations

$$\text{Planned run time} \quad a = 100 \text{ hours}$$
$$\text{Actual run time} \quad b = 90 \text{ hours}$$

(owing to breakdowns, setups)

$$\text{Expected output in actual run time} \quad c = 1000 \text{ units in the 90 hours}$$
$$\text{Actual output} \quad d = 900 \text{ units}$$

(owing to reduced speed, minor stoppages)

$$\text{Expected quality output} \quad e = 900 \text{ units}$$
$$\text{Actual quality output} \quad f = 800 \text{ units}$$

(owing to scrap, rework, startup losses)

$$\text{OEE} = \frac{b}{a} \times \frac{d}{c} \times \frac{f}{e} = \frac{90}{100} \times \frac{900}{1000} \times \frac{800}{900} = 72\%$$

*OEE calculation for an automated press line*

### Working pattern

- Three shifts of 8 hours, 5 days per week.
- Tea breaks of 24 minutes per shift.

### Data for week

- 15 breakdown events totalling 43 hours.
- 5 die changes averaging 4 hours each per setup and changeover.
- 15 500 units produced, plus 80 units scrapped, plus 150 units requiring rework.
- Allowed time as planned and issued by production control for the 5 jobs was 52 hours including 15 hours for setup and changeover.

## OEE for week

Loading time = attendance − tea breaks = 120 − 6 = 114 hours

Downtime = breakdowns + setups and changeovers = 43 + 20 = 63 hours

$$\text{Availability} = \frac{114 - 63}{114} = 44.7\%$$

Actual press running time (uptime) = 120 − 6 − 43 − 20 = 51 hours

Allowed press running time = 52 − 15 = 37 hours

$$\text{Performance efficiency} = \frac{37}{51} = 72.5\%$$

Product input (units) = 15 500 + 80 + 150 = 15 730

Quality (first time) product output (units) = 15 500

$$\text{Quality rate} = \frac{15\,500}{15\,730} = 98.5\%$$

$$\text{OEE} = 0.447 \times 0.725 \times 0.985 = 31.9\%$$

## Data for four-week period

Over a recent four-week period the following OEE results were obtained:

| Week | OEE (%) | = | Availability (%) | × | Performance rate (%) | × | Quality rate (%) |
|------|---------|---|------------------|---|----------------------|---|------------------|
| 1 | 44.6 | = | 65.0 | × | 70.0 | × | 98.0 |
| 2 | 43.8 | = | 58.0 | × | 77.0 | × | 98.0 |
| 3 | 36.7 | = | 47.0 | × | 80.0 | × | 97.5 |
| 4 | 31.9 | = | 44.7 | × | 72.5 | × | 98.5 |
| Average | 39.4 | = | 53.7 | × | 74.9 | × | 98.0 |

## Best of best OEE and potential benefit

The best of best OEE can now be calculated. In addition, if the hourly rate of added value is taken to be £100, the annual benefit (45 week year) of moving from the current average OEE of 39.4% to the best of best can be found.

Best of best OEE = availability × performance × quality

$$= 65.0 \times 80.0 \times 98.5 = 51.2\%$$

Potential loading hours per year = 114 × 45 = 5130

At 39.4% OEE, value added per year = 0.394 × 5130 × £100 = £202,122

At 51.2% OEE, value added per year = 0.512 × 5130 × £100 = £262,656

Therefore a benefit of £60,534 is possible by consistently achieving best of best through tackling six losses using the TPM improvement plan.

## 5.3  Improvement cycle

### *Step 7: Assessment of the six big losses*

The importance of understanding and tackling the six big losses cannot be over-emphasized! They were listed in Chapter 3 and illustrated by the iceberg analogy in Figure 3.4, repeated here as Figure 5.20. The six losses are:

- breakdowns
- setup and adjustment
- idling and minor stoppages
- speed
- quality defect and rework
- startup.

These are elaborated in Figures 5.21 to 5.26. The relationship of these losses to the OEE is set out in Figure 5.18.

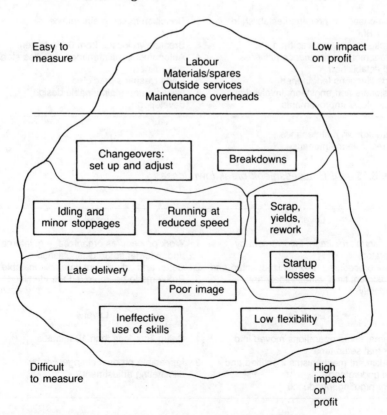

**Figure 5.20**  *True cost of maintenance: seven-eighths hidden*

We will develop a detailed definition in later chapters regarding the four levels of control referred to under each of the six losses in Figures 5.21–5.26. However, in order to give an early indication a definition is as follows:

*Level 1*   Introductory phase of TPM with pilots: 3 to 6 months on.
*Level 2*   Refine best practice and standardize: 6 to 12 months on.
*Level 3*   Build capability: 12 to 18 months on.
*Level 4*   Continuous improvement: 18 to 24 months and beyond.

The improvement cycle in TPM starts from an appreciation of what the six big losses are and proceeds through problem solving to the establishment of

| Level 1 | Level 2 |
|---|---|
| 1 Combination of sporadic and chronic breakdown | 1 Chronic breakdowns |
| 2 Significant breakdown losses | 2 Breakdown losses still significant |
| 3 BM > PM | 3 PM = BM |
| 4 No operator asset care | 4 Operator asset care implemented |
| 5 Unstable lifespans | 5 Parts lifespans estimated |
| 6 Equipment weaknesses not recognized | 6 Equipment weaknesses well acknowledged |
| | 7 Maintainability improvement applied on above points |

| Level 3 | Level 4 |
|---|---|
| 1 Time-based maintenance established | 1 Condition-based maintenance |
| 2 PM > BM | 2 PM |
| 3 Breakdown losses less than 1% | 3 Breakdown losses from 0.1% to zero |
| 4 Autonomous maintenance activities well established | 4 Autonomous maintenance activities stable and refined |
| 5 Parts lifespans lengthened | 5 Parts lifespans predicted |
| 6 Designers and engineers involved in higher-level improvements | 6 Reliable and maintainable design developed |

BM   breakdown maintenance
PM   predictive maintenance

**Figure 5.21**   *OEE assessment: breakdown losses*

| Level 1 | Level 2 |
|---|---|
| 1 No control: minimum involvement by operators | 1 Work procedures organized, e.g. internal and external setup distinguished |
| 2 Work procedures disorganized: setup and adjustment time varies widely and randomly | 2 Setup and adjustment time still unstable |
| | 3 Problems to be improved are identified |

| Level 3 | Level 4 |
|---|---|
| 1 Internal setup operations moved into external setup time | 1 Setup time less than 10 minutes |
| 2 Adjustment mechanisms identified and well understood | 2 Immediate product changeover by eliminating adjustment |
| 3 Error proofing introduced | |

**Figure 5.22**   *OEE assessment: setup and adjustment losses*

| Level 1 | Level 2 |
|---|---|
| 1 Losses from minor stoppages unrecognized and unrecorded | 1 Minor stoppage losses analysed quantitatively by: frequency and location of occurrence; volume lost |
| 2 Unstable operating conditions due to fluctuation in frequency and location of losses | 2 Losses categorized and analysed; preventive measures taken on trial-and-error basis |

| Level 3 | Level 4 |
|---|---|
| 1 All causes of minor stoppages analysed: all solutions implemented | 1 Zero minor stoppages (unmanned operation possible) |

**Figure 5.23** *OEE assessment: idling and minor stoppage losses*

| Level 1 | Level 2 |
|---|---|
| 1 Equipment specifications not well understood | 1 Problems related to speed losses analysed: mechanical problems, quality problems |
| 2 No speed standards (by product and machinery) | 2 Tentative speed standards set and maintained by product |
| 3 Wide speed variations across shifts/operators | 3 Speeds vary slightly |

| Level 3 | Level 4 |
|---|---|
| 1 Necessary improvements being implemented | 1 Operation speed increased to design speed or beyond through equipment improvements |
| 2 Speed is set by the product. Cause-and-effect relationship between the problem and the precision of the equipment | 2 Final speed standards set and maintained by product |
| 3 Small speed losses | 3 Zero speed losses |

**Figure 5.24** *OEE assessment: speed losses*

| Level 1 | Level 2 |
|---|---|
| 1 Chronic quality defect problems are neglected | 1 Chronic quality problems quantified by: details of defect, frequency; volume lost |
| 2 Many reactive and unsuccessful remedial actions have been taken | 2 Losses categorized and reasons explained: preventive measures taken on trial-and-error basis |

| Level 3 | Level 4 |
|---|---|
| 1 All causes of chronic quality defects analysed: all solutions implemented, conditions favourable | 1 Quality losses from 0.1% to zero |
| 2 Automatic in-process detection of defects under study | |

**Figure 5.25** *OEE assessment: quality defect and rework losses*

| Level 1 | | Level 2 | |
|---|---|---|---|
| 1 | Startup losses not recognized, understood or recorded | 1 | Startup losses understood in terms of breakdowns and changeovers |
| | | 2 | Startup losses quantified and measured |
| Level 3 | | Level 4 | |
| 1 | Process stabilization dynamics understood and improvements implemented | 1 | Startup losses minimized through process control |
| 2 | Causes due to minor stops aligned with startup losses | 2 | Remedial actions on breakdowns, setups, minor stops and idling minimize startup losses |

**Figure 5.26**   *OEE assessment: startup losses*

best practice routines. Finding the root cause of the six losses is developed in step 8 of the TPM improvement plan.

### Step 8: Problem solving

#### P-M analysis

In seeking to solve the problems which lie behind the six big losses, TPM uses P-M analysis, introduced in Chapter 3. This emphasizes the machine/ human interface: there are *phenomena* which are *physical*, which cause *problems* which can be *prevented*; these are to do with *materials, machines, mechanisms* and *manpower*.

These problems may have a single cause, multiple causes or a complex combination of causes. P-M analysis is concerned with pin-pointing the causes, taking counter-measures and evolving best practice routines so that the problems are dealt with once and for all.

#### On-the-job reality

This approach recognizes four principles as follows. Each principle is followed by practical on-the-job steps.

*Cleaning is inspection*
- Operators are encouraged to look for opportunities to reduce accelerated deterioration and improve equipment design.

*Detect problems and opportunities*
- Work with the team to systematically review problems and opportunities to achieve target performance.

*Restore and then improve*
- Restore before renew: that is, solve existing problems before introducing new equipment with new problems.

- Make all aware of the problem and the opportunity.
- Observe the current situation.
- Define the problem and the conditions under which it occurs.
- Develop the optimum solution.
- Try out new ideas first and check the results.
- Apply proven low-cost or no-cost solutions first.
- Implement ideas as soon as possible.
- Standardize best practice with all those involved.
- Monitor and review.

*Recognize and reward effort*
- Supervisors create the expectation of pride in the workplace.
- Encourage a positive attitude to problem detection and long-term elimination.
- Focus on hidden costs as well as the more visible production output, using performance trends for OEE, number of faults eliminated, value of OEE losses, percentage maintenance quality.

The first task is to classify reliability problems under the headings corresponding to the six losses, and to use the P-M problem solving approach to identify and then eliminate the possible causes of the phenomena.
    The subsequent steps are:

1   Formulate improvement plans which are no cost or low cost and are therefore also low risk.
2   Investigate malfunctions caused by technical problems and propose improvements which will involve some cost and some risk.
3   Investigate support problems and make proposals to overcome them, i.e.
      - support services: parts handling, spare parts procurement
      - support equipment: tool change equipment, maintenance tooling.

Some simple improvements are highlighted in Figure 5.27, and Figure 5.28 emphasizes the importance of setting targets and having standards.

Maintainability
- Standard fixtures
- Easy to remove covers
- Right tools for the job
- Local instruments

Accessibility
- CAN-DO
- Standard heights
- Thoughtful guarding
- Planned installation

**Figure 5.27**   *Some simple improvements*

---

There can be no improvements where there are no standards:
- Analyse data
- Convert to information
- Set targets and standards
- Measure benefit potential
- Manage the necessary changes

---

**Figure 5.28**   *Importance of standards in improvement*

### The problem solving circle

The problem solving process is further illustrated by Figure 5.29. This approach is elaborated as follows:

1  *Identify*
   - Define the problem.
2  *Analyse*
   - Do a physical analysis of the problem.
   - Isolate every condition that might cause the problem.
   - Evaluate equipment, material, method and environment.
   - Plan the investigation.
   - Investigate malfunctions.
   - Generate possible solutions.
3  *Plan*
   - List the tasks to be achieved over a reasonable period.
   - Identify the best sequence including training plans.
   - Identify responsibilities and resources.

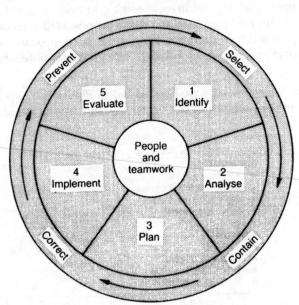

**Figure 5.29**   *Problem solving circle*

- Produce an outline timetable together with resources and objectives for time, cost and quality of implementation.
- Formulate improvement plans.
- Present plans.

4 *Implement*
5 *Evaluate*
- Check and take action as required.

## The problem solving cascade

This is illustrated in Figure 4.5, repeated here as Figure 5.30. A typical TPM problem solving document is shown in Figure 5.31.

## Equipment consciousness

An essential prerequisite for a successful approach to problem solving, reduction of the six losses and establishment of best practice routines lies

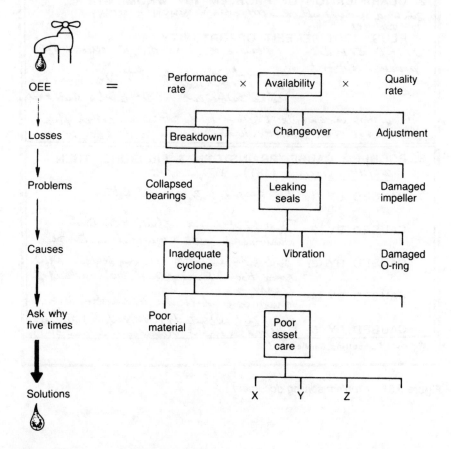

**Figure 5.30** *Problem solving cascade*

Issue 4   4/1/92

# TPM PROBLEM SOLVING DOCUMENT

MACHINE / PART NAME: 10HP Motor - Casters   TEAM LEADER: Ron Davies

PROBLEM RAISED BY: J Smith   DATE: 21/11/91

### 1. PROBLEM STATEMENT (SPECIFIC)

10 HP Motor AB23/106 A which Drives Water Pump for Caster-line cooling keeps burning out (4 times since 1/8/91)

### 2. CLARIFICATION OF PROBLEM (BY WHOM, WHERE, WHEN & HOW)

Jim Smith Discussed with RD (TM Casters) on 24/11/91

### PLUS COST/BENEFIT OPPORTUNITY

JS/RD and Phil Davis (M'tenance Elec) on-the-job session 27/11/91 problem-solving.

1. Burn out because of stressing?
2. Try 15 HP Motor?
3 Maybe Dust & Dirt Ingress therefore shield Motor?

N.B. 4 stops in 3 months 6½ Hours total D/time on casters plus £200 L&M to fix = £6,500 Lost Added Value = c. £26,800 per year

### 3. PROBLEM CAUSE (BRAINSTORM & FISHBONE, THEN 27/11/91 Session LIST) RD/JS/PD

CAUSED BY 1: Motor Stressing : Bigger HP Rating

CAUSED BY 2: Dust & Dirt Ingress : Shield Motor Bearing What about other equipment in the Area?

CAUSED BY 3: Dust Extraction U/S : Wrong position. Need extra Hood. Source of Dust Build up : Lack of cleaning

CAUSED BY 4: Pump over-stressed : Recondition seals.

CAUSED BY 5: Open Roof Water Tank (50K gallons) Full of Algae. Filter Blockage Vacuum Blow Ball

Willmott Consulting Services

**Figure 5.31**   *Problem solving document*

## TPM PROBLEM SOLVING DOCUMENT

**4. CLARIFICATION OF ROOT CAUSES    (DO THE CAUSES**
*5/12/91 Team Appraisal. Session* **EXPLAIN THE**
**PROBLEM?)**

- Dust & Dirt Ingress to Motor + others(?)
- Pump Recondition Needed urgently
- Roof Tank Algea very problem leads to block filter causing pump vacuum causing pump & Motor stops.

**5. COUNTER-MEASURE    (ACTIONS REQUIRED TO**
**TEMPORARY:        RESOLVE CAUSE(S))**

1) shield motor & pump
2) re-seal pump                          } completed
3) unblock & clean roof tank filters  }  10/12/91

**PERMANENT:**
4) Improve dust extraction via addition Hood
5) call in Water Authority chemist re Algea removal
6) Regular Roof inspection re tank (repair fortnightly- winter)
(per week summer)

**6. CONFIRMATION OF COUNTER-MEASURE  (HAVE ACTIONS**
4) Planned for April Shutdown              **CLEARED**
5) WA Repair received 22/12/91             **PROBLEM)**
6) start checks agreed w/c 4/1/92

**7. FEEDBACK  (WHO ELSE NEEDS TO KNOW)**
- Send copy to Scunthorpe Works TPM facilitator
- Nominate JS and PD for TPM Excellence Award (January '92)

Willmott Consulting Services

**Figure 5.31**  *(continued)*

in training operators to be equipment conscious. Some examples, checklists and techniques are given below.

## Recurring breakdown problems

Overheating, vibration and leakage are problems which will constantly arise and, unless tackled and eliminated once and for all, will continue to contribute to breakdown losses. Figures 5.32 to 5.35 offer approaches to these problems.

## A structured approach to setup reduction

Figure 5.36 draws attention to all the points which must be looked at and evaluated. An indication of the importance of tackling adjustment is given

| Cause | Remedy |
|---|---|
| 1 Excessive vibration | Cure cause |
| 2 Unabosrbent mountings | Refit new mountings |
| 3 Insufficient mountings or supports | Fit extra |
| 4 Wrong grade/type component fitted | Fit correct grade |
| 5 Poor fitting | Refit correctly |
| 6 Overheating | Seek and cure cause |
| 7 Technical ignorance/innocence | Retrain |
| 8 Material breakdown | Replace |

Vibration is one of the major causes of fittings or fixings working loose and giving rise to leaks. Other items contribute, such as poor fitting, or overheating which causes seals first to bake and then to crack.

To identify leaks:
- In the case of liquids: puddles will form.
- In the case of gases: noise, smell or bubbles when tested with soapy water.

**Figure 5.32** *Problem solving: leakages*

| Cause | Remedy |
|---|---|
| 1 Excessive lubrication | Remove excess |
| 2 Incorrect lubricant | Replace with correct |
| 3 Lubrication failure/contimination | Check cause and remedy |
| 4 Low lubricant level | Top up |
| 5 Poor fitting | Refit correctly |
| 6 Excessive speed above standard | Reduce speed to standard |
| 7 Overloading | Reduce loading |
| 8 Blockages in system | Clean and flush system |
| 9 Excessive pipe lengths or joints | Redesign system |

**Figure 5.33** *Problem solving: overheating*

When overheating can be attributed to a lubricating problem, it is always best policy to remove all lubricant and replace with new after the problem has been cured.

Lubricant which has overheated starts to break down and will not perform as it should.

Identification of overheating:

*Visual*  Items that have overheated will discolour or give off smoke.
*Smell*  Overheated items in many cases give off fumes which can be smelt.
*Touch*  By touching items suspected of overheating one can tell, but caution must be exercised in the first instance. A hand held close to the item will indicate whether it would burn if touched.
*Electrical/visual*  Many items of equipment have built-in temperature sensing devices and these should be monitored regularly. An awareness of the significance of the temperature readings is essential.

**Figure 5.34**  *Problem solving: overheating and lubrication*

| | Cause | Remedy |
|---|---|---|
| 1 | Out of balance | Correct or repflace |
| 2 | Bent shafts | Straighten or replace |
| 3 | Poor surface finish | Rework surface |
| 4 | Loose nuts and bolts | Tighten |
| 5 | Insecure clips | Secure clips |
| 6 | Insufficient mountings | Get extra added |
| 7 | Too rigid mountings | Get softer ones |
| 8 | Slip stick | Lubricate |
| 9 | Incorrect grade lubricant | Clean and replace |
| 10 | Worn bearings | Replace |
| 11 | Excessive speed above standard | Reduce speed to standard |

Some of the remedies will require a skilled maintenance fitter. Some can be carried out by the operator with some training: items 4, 5, 8, 9 and 11.

Vibration is identified by sight, touch or noise increase.

**Figure 5.35**  *Problem solving: vibration*

by the percentage figures based on hard experience and shown in Figure 5.37.

Setup and adjustment are so important in the drive towards reduced losses, better equipment effectiveness and ultimately world class manufacture. Shigeo Shingo, the guru of Single-Minute Exchange Die (SMED), states the following in his book *A Revolution in Manufacturing: the SMED System*:

● Every machine setup can be reduced by 75%.

What a challenge for Western companies! The SMED approach uses a derivative of the Deming circle:

*Focus*  Setup video.
*Analyse*  Pareto, ergonomics.
*Develop*  Script, simulate, agree.
*Execute*  Train, measure, honour, empower.

| **External setup** | | |
|---|---|---|
| Preparations | • Tools (types, quantities)<br>• Locations<br>• Position<br>• Workplace organization and housekeeping<br>• Preparation procedure | • Don't search<br>• Don't move<br>• Don't use |
| Preparation of ancillary equipment | • Check jigs<br>• Measuring instruments<br>• Preheating dies<br>• Presetting | |
| **Internal Setup** | | |
| Operation phase | • Standardize work procedures and methods<br>• Allocate work<br>• Evaluate effectiveness of work<br>• Parallel operations<br>• Simplify work<br>• Number of personnel<br>• Simplify assembly<br>• Assembly/integration<br>• Elimination | • Eliminate redundant procedures<br>• Reduce basic operations |
| Dies and jigs | • Clamping methods<br>• Reduce number of clamping parts<br>• Shapes of dies and jigs: consider mechanisms<br>• Use intermediary jigs<br>• Standardize dies and jigs<br>• Use common dies and jigs<br>• Weight<br>• Separate functions and methods<br>• Interchangeability | • Make it easy |
| Adjustment | • Precision of jigs<br>• Precision of equipment<br>• Set reference surfaces<br>• Measurement methods<br>• Simplification methods<br>• Standardize adjustment procedures<br>• Quantification<br>• Selection<br>• Use gauges<br>• Separate out interdependent adjustments<br>• Optimize conditions | • Eliminate adjustment |

**Figure 5.36**   *Factors in reduction of setup and adjustment time*

| | |
|---|---|
| Preparation of materials, jigs, tools and fittings | 20% |
| Removal and attachment of jigs, tools and dies | 20% |
| Centring, dimensioning | 10% |
| Trial processing, adjustment | 50% |

**Figure 5.37**   *Adjustments as a percentage of total setup time*

In the SMED system, the success is subject to some conditions:

*An attitude*   The team wants to score.
*An empowerment*   The team has a budget.
*An involvement*   Management is part of the team.
*A commitment*   Management sets the target.
*A philosophy*   Step-by-step improvement.

Moreover, the SMED approach suggests that there are characteristics common to all setups:

- Prepare, position, adjust, store away.
- Internal and external activities.
- From last good product to first good product.

Figure 5.38 shows the three steps towards a cumulative reduction of 75% to 95% in setup time in the SMED system. A graphical representation of the reductions achieved is shown in Figure 5.39.
   TPM develops six conceptual steps for analysing adjustment operations.

*Purpose*   What function is apparently served by adjustment?
*Current rationale*   Why is adjustment needed at present?
*Method*   How is the adjustment performed?
*Principles*   What is the true function of the adjustment operation as a whole?
*Causal factors*   What conditions create the need for adjustment?
*Alternatives*   What improvements will eliminate the need for adjustment?

Figure 5.40 provides a clear visual presentation of the TPM approach to analysing adjustment operations leading to minimization of losses. Figure 5.41 reviews progressively the process from an analysis of the present

Step 1: separate internal/external activities

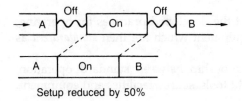

Setup reduced by 50%

Step 2: shorten internal activities

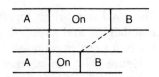

Setup reduced another 50%

Step 3: minimize external activities and continue reducing internal tasks

Cumulative reduction between 75% and 95%!

**Figure 5.38**   *SMED steps to reducing setup time*

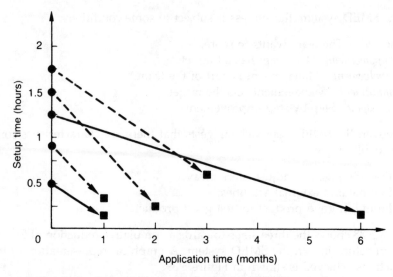

**Figure 5.39**   *Setup times reduced significantly by SMED approach*

position right through to achieving optimal conditions. Wherever possible make use of video: it is a very powerful analysis tool.

### Eliminating reasons for speed losses

Figure 5.42 provides a decision-free structure to help eliminate reasons for running at reduced speeds. Figure 5.43 provides a checklist of ideas for developing approaches to increase speeds.

### Step 9: Best practice routines

This final step brings together all of the developed practices for operating, maintaining and supporting the equipment which are then *standardized* as the best practice routines.

Figure 5.44 summarizes the relationship between standard operation, techniques for asset care and the right tools, spares, facilities and equipment.

Standard operation ensures:

- reduced chance of error and risk
- removal of performance irregularity
- elimination of poor operation as a cause of problems
- simplified training within and between shifts.

When launching a pilot, we should consider:

- What is the best method of operating our pilot?
- Does the team agree?
- Does each shift agree?
- Do the key contacts agree?

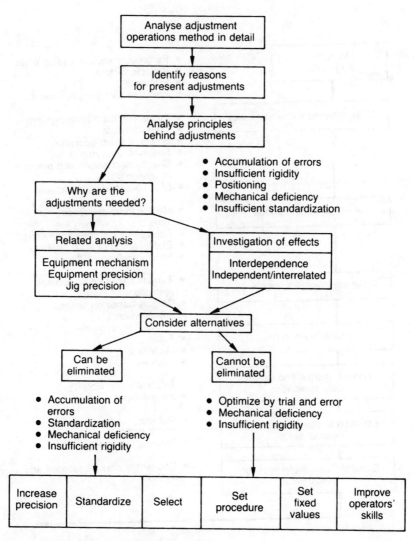

**Figure 5.40**  *Analysis of adjustment operations*

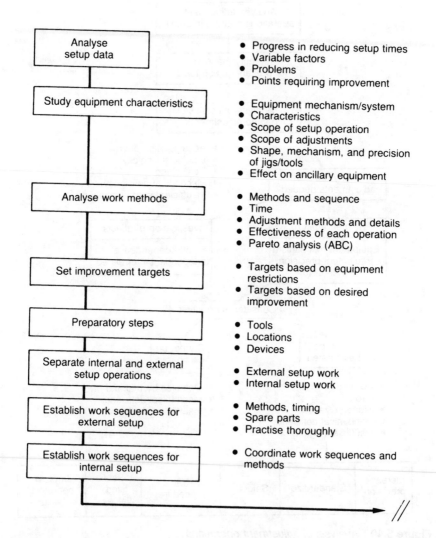

**Figure 5.41** *Process of improving setup and adjustment*

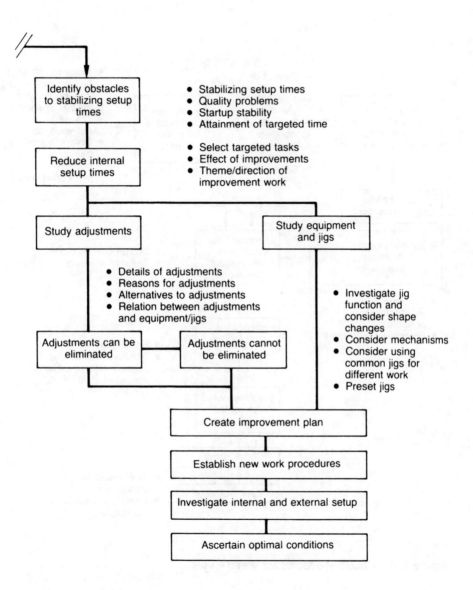

**Figure 5.41**  *(Continued)*

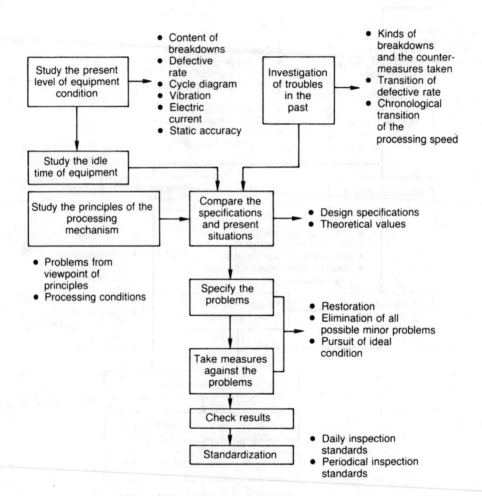

**Figure 5.42**   *Counter-measures for speed losses*

| Determine present levels | • Speed<br>• Bottleneck processes<br>• Downtime, frequency of stoppages<br>• Conditions producing defects |
| --- | --- |
| Check differences between specification and present situation | • What are the specifications?<br>• Difference between standard speed and present speed<br>• Difference in speeds for different products |
| Investigate past problems | • Has the speed ever been increased?<br>• Types of problems<br>• Measures taken to deal with past problems<br>• Trends in defect ratios<br>• Trends in speeds over time<br>• Differences in similar equipment |
| Investigate processing theories and principles | • Problems related to processing theories and principles<br>• Machining conditions<br>• Processing conditions<br>• Theoretical values |
| Investigate mechanisms | • Mechanisms<br>• Rated output and load ratio<br>• Investigate stress<br>• Revolving parts<br>• Investigate specification of each part |
| Investigate present situation | • Processing time per operation (cycle diagram)<br>• Loss times (idling times)<br>• Check precision of each part<br>• Check using five senses |

**Figure 5.43**  *Strategies for increasing speed*

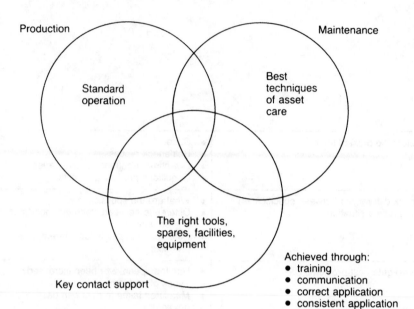

**Figure 5.44**  *Best practice*

- How do we train people to operate this method? (*single-point lessons*)
- How do we communicate this method to each shift? (*visual management*)
- How do we make it easy to do it right and difficult to do it wrong? (*improvements*)

For each piece of equipment we need to establish the best practice for:

- provision of tools
- provision of spares
- monitoring instruments
- outside contracts
- warranties
- technical help.

We must therefore involve the key contacts.

In effect the best practice routine is similar to your motor car handbook. It explains the best and correct way to operate, maintain and support the car. It gives the standard operation and asset care procedures.

# Applying the TPM Improvement Plan

## 6.1 Training context

The following example is based on a TPM improvement plan training exercise as part of a four-day TPM facilitator training course. Approximately 70% of the course content is putting the theory of TPM into practice on a *live* TPM pilot piece of equipment. This practical focus is so that the facilitators are experiencing over four days what they will be coaching their own TPM teams over the twelve to twenty weeks of a TPM pilot project, the process of which is described in later chapters.

The output of this particular exercise is based on a one-hour presentation which the five budding TPM facilitators made after spending two and a half days assimilating and using the nine-step TPM improvement plan. The following sections 6.2 to 6.16 inclusive are the content of their presentation.

## 6.2 Team brief

A core team is undertaking a pilot TPM project. The team is made up of:

- three production personnel (one per shift)
- two maintenance personnel (one electrical, one mechanical).

The company – Merlin Gun Technology – is planning to introduce TPM across the site (200 personnel).

The pilot aims to develop a practical, model example of equipment operating under TPM. This will support the roll-out of TPM. It will also highlight those issues which need to be overcome to achieve a successful implementation. Specifically, the team will:

- assess the critical elements of the equipment
- identify what refurbishment is required to put the equipment into good condition
- develop an asset care and history recording process
- identify the main problem areas and develop solutions
- establish the level of equipment effectiveness and set targets for improvement
- produce an implementation plan to improve the equipment reliability.

## 6.3   Company Information

### Background

Merlin Gun Technology has the following characteristics:

- The company makes welding guns and welding tips.
- Most of the volume is in welding tips.
- The company experiences pressure from customers to produce in small batches.
- The company is expanding into further export markets.
- Department 50 is recognized as the main bottleneck.

The company organization is shown in Figure 6.1.

### Department 50

Department 50 makes the most popular welding tips. It has always been the bottleneck department.

The department produces basically three types of tip (Figure 6.2):

- the 5020, a flat-head tip produced in two operations
- the 5031, a tip with one angled face produced by three operations
- the 5042, a pointed tip with two angled faces produced in four operations.

There are three machines:

- the L101 computer numerically controlled (CNC) lathe

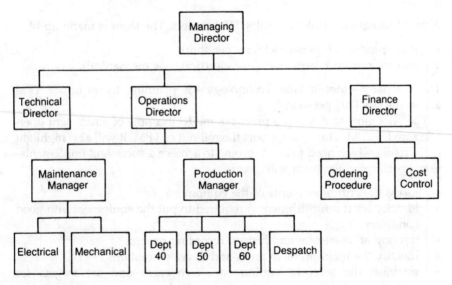

**Figure 6.1**   *Merlin Gun Technology: organization*

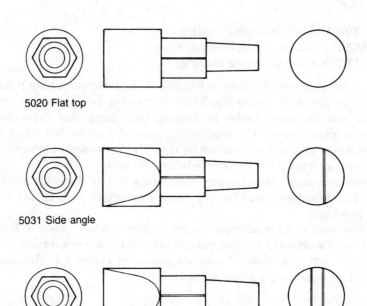

5020 Flat top

5031 Side angle

5042 Pointed

**Figure 6.2**   *Department 50 products*

- the M201 CNC Bridgeport Interact milling machine
- the M202 Denford Easimill 3 milling machine.

The machine operating data are shown in Figure 6.3. It should be noted that the times have been developed by the planners based on experience. The machines should be capable of the following cycle times:

| Part | Forecast | Operations | | | | Total cycle time (mins) |
|------|----------|------|------|------|------|------|
| | | L101 | M201 | M202 | | |
| | | | | RH angle | LH angle | |
| 5020 | 690 | ✓ | ✓ | | | 13.50 |
| 5031 | 200 | ✓ | ✓ | ✓ | | 29.50 |
| 5042 | 65 | ✓ | ✓ | ✓ | ✓ | 45.50 |
| Output per week | | 955 | 850 | 330 | | |
| Scheduled running time (min) | | 7200 | 4800 | 5200 | | |
| Scheduled time per cycle (min) | | 7.5 | 6.00 | 16.00 | 16.00 | |

**Figure 6.3**   *Department 50 machine operating data*

- L101: 4.12 min including loading
- M201: 3 min including loading
- M202: 8.5 min including loading.

As can be seen from the data in Figure 6.3, the lathe is running three shifts, five days per week; the miller M201 is running two shifts, five days per week; and the other miller is running two shifts plus overtime. Some cover is provided for the machines during the shifts, but there are still times when the machines cannot be run. These planned stoppages are one of the reasons for the difference between the two.

There are some tool changes. Sometimes this is done by maintenance outside production time. Usually some production time is lost owing to tool breakages.

There seem to be occasional quality problems where much of the output has to be reworked to remove burrs. Mostly there are few quality problems.

Department 50's financial data are shown in Figure 6.4. The company's shop floor logistics are given in Figure 6.5.

## 6.4   TPM presentation and plan

The equipment chosen for the pilot project is the M201 milling machine.
The TPM presentation for the M201 team had the following headings:

- introduction
- plan
- equipment description
- criticality assessment
- condition appraisal
- refurbishment
- asset care
- equipment history
- OEE assessment
- the six losses
- problems/improvements
- best practice
- implementation plan
- concluding remarks.

The schedule for the project is shown in Figure 6.6.

As discussed in earlier chapters, the TPM improvement plan has the following three cycles and nine steps:

*Condition cycle*
1   Critically assess the equipment.
2   Carry out an appraisal of its condition.
3   Decide on the refurbishment programme.
4   Determine the future asset care regime.
*Measurement cycle*
5   Decide what to record and monitor.

**Profit forecast**

| | £000 | £000 |
|---|---|---|
| Sales | | 1322.36 |
| | | |
| Materials | 550 | |
| Consumables | 2.72 | |
| Inventory adjustment | 6.13 | |
| Material costs | | 558.85 |
| | | |
| Direct labour | 138.15 | |
| Pensions | 23.14 | |
| Holiday pay | 13.27 | |
| Labour costs | | 174.56 |
| | | |
| Production | 188.55 | |
| Depreciation | 37.78 | |
| Production overheads | | 226.33 |
| | | |
| Technical | 61.26 | |
| Administration | 104.15 | |
| Selling | 77.26 | |
| Finance | 33.35 | |
| Other overheads | | 276.02 | 22% |
| | | |
| Total costs | | 1235.76 |
| | | |
| Profit | | 86.60 | 7% |

**Cost apportionment**

| Part | Production forecast | Material cost/piece (£) | Cycle time (min) | Machine time (h) | Fixed costs/ product (£) | Fixed costs/ piece (£) |
|---|---|---|---|---|---|---|
| 5020 | 45 540 | 9.40 | 13.5 | 10 246.50 | 228 046 | 5.01 |
| 5031 | 10 120 | 9.40 | 29.5 | 4 975.67 | 110 739 | 10.94 |
| 5042 | 3 680 | 9.40 | 45.5 | 2 790.67 | 62 109 | 16.88 |
| | 59 340 | | | 18 012.84 | 400 894 | |

**Revenue forecast**

| Part | Total cost/ piece (£) | Operational margin (%) | Selling price (£) | Total revenue (£) |
|---|---|---|---|---|
| 5020 | 14.41 | 25 | 18.01 | 820 153 |
| 5031 | 20.34 | 50 | 30.51 | 308 800 |
| 5042 | 26.28 | 100 | 52.55 | 193 402 |
| | | | | 1 322 355 |

**Figure 6.4** *Department 50 financial data*

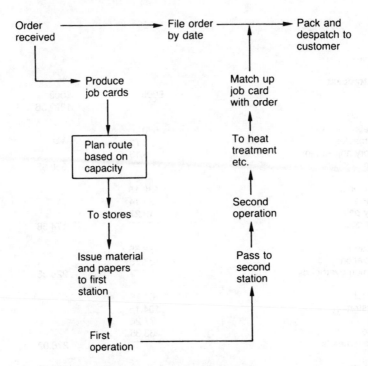

**Figure 6.5**   *Merlin Gun Technology: shop floor logistics*

| Activity | Tuesday p.m. | Wednesday a.m. | Wednesday p.m. | Thursday a.m. |
|---|---|---|---|---|
| Plan | | | | |
| Equipment description | | | | |
| Criticality assessment | | | | |
| Condition appraisal | | | | |
| Refurbishment plan | | | | |
| Asset care | | | | |
| Equipment history | | | | |
| OEE assessment | | | | |
| The six losses | | | | |
| Problems/improvements | | | | |
| Best practice | | | | |
| Implementation plan | | | | |
| Presentation preparation | | | | |

**Figure 6.6**   *Miller M201 project management plan*

6   Decide the OEE measures: best of best, world class.
*Improvement cycle*
7   Assess the six losses.
8   Achieve improvement through problem solving.
9   Agree on best practice routines.

The plan flow is shown again here as Figure 6.7.

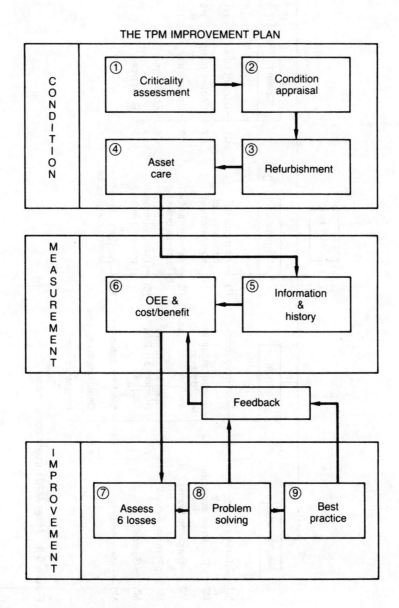

**Figure 6.7**   *TPM improvement plan sequence*

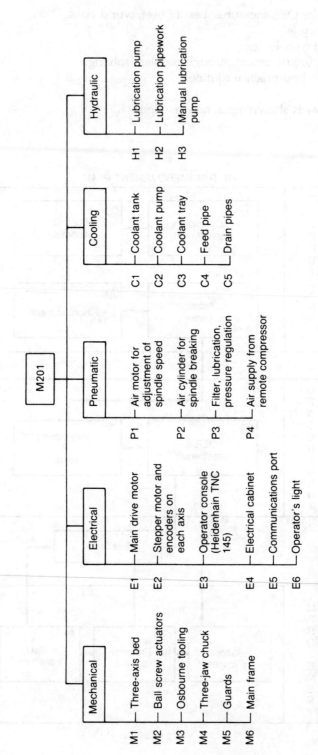

**Figure 6.8** *Miller M201 major components*

## 6.5   Equipment description

The components of the M201 miller are described in Figure 6.8 and illustrated in Figure 6.9.

The performance data for the M201 are given in Figure 6.10. An equipment history is provided by Figure 6.11.

The M201 operation cycle is shown in Figure 6.12. The operations layout for Department 50 is given in Figure 6.13.

## 6.6   Criticality assessment

The main systems and the components of M201 are as follows:

*Electrical*
- supply panel
- isolators
- motors
- control unit
- lighting
- cables

*Hydraulics/lubrication*
- pump
- pipework

*Air supply*
- compressor
- regulator
- lubricator
- moisture trap
- pipework

*Coolant*
- pump
- sump
- flexible nozzle
- pipework

*Mechanical*
- spindle unit
- table traverse
- ball screw
- tool chuck
- work chuck
- guards
- tool bit

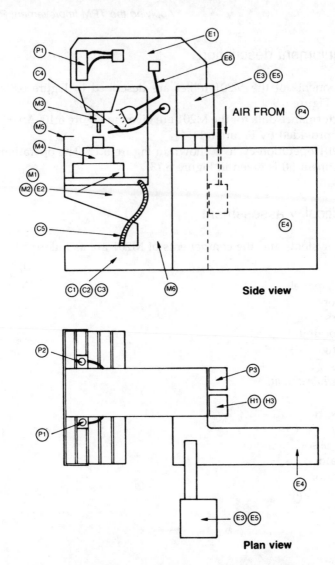

**Figure 6.9**  *Miller M201 component diagram: see Figure 6.8 for key*

| | |
|---|---|
| Maximum spindle speed | 3750 rpm |
| Maximum axis feed rate | 1 m/min |
| | |
| Current spindle speed | 2000 rpm |
| Current x and y axes feed rates | 0.5 m/min |
| Current z axis feed rate | 25 mm/min |
| | |
| Time for first cut of hexagonal | 33 s |
| Time for subsequent hexagonal cuts | 15 s each |
| Total number of subsequent cuts | 4 |
| Total time for all cuts | 93 s |
| | |
| z axis travel per cut | 2 mm |
| z axis total travel | 10 mm |

**Figure 6.10**  *Miller M201 performance data*

Cycle time ____3.00____ min

| Week no. | Day, based on 1440 min | Planned stoppages (min) | Planned availability (min) | Unplanned stoppages (min) | Changeover (min) | Uptime (min) | Completed cycles | Rework | First time OK |
|---|---|---|---|---|---|---|---|---|---|
| 15 | M | 530 | 910 | 120 | | 790 | 187 | | 187 |
| | T | 530 | 910 | 153 | 20 | 737 | 185 | | 185 |
| | W | 530 | 910 | 96 | | 814 | 124 | 13 | 111 |
| | Th | 530 | 910 | 132 | 20 | 759 | 209 | | 209 |
| 16 | F | 530 | 910 | 129 | 13 | 769 | 175 | 13 | 162 |
| | M | 530 | 910 | 90 | | 820 | 151 | 118 | 33 |
| | T | 530 | 910 | 42 | 9 | 859 | 228 | | 228 |
| | W | 530 | 910 | 42 | 9 | 859 | 228 | | 228 |
| | Th | 530 | 910 | 99 | 20 | 791 | 121 | | 121 |
| 17 | F | 530 | 910 | 50 | | 860 | 201 | | 201 |
| | M | 530 | 910 | 27 | 32 | 851 | 223 | | 223 |
| | T | 530 | 910 | 115 | 12 | 783 | 203 | 51 | 152 |
| | W | 530 | 910 | 34 | 24 | 852 | 214 | | 214 |
| | Th | 380 | 1060 | 356 | 5 | 700 | 132 | | 132 |
| | F | 530 | 910 | 46 | 17 | 847 | 236 | 76 | 160 |

Figure 6.11  Miller M201 equipment history

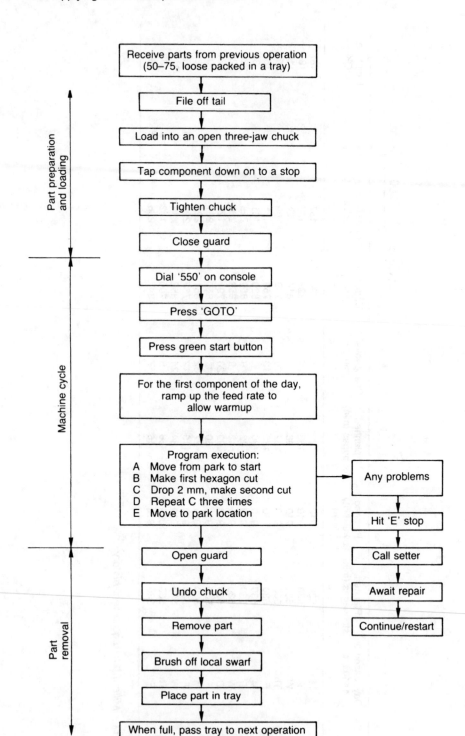

**Figure 6.12**   *Miller M201 machine/operator cycle*

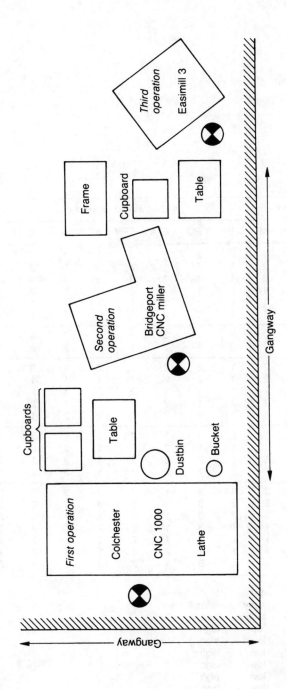

**Figure 6.13** *Department 50 operations layout*

The criticality assessment matrix for these components is given in Figure 6.14. From this matrix the following conclusions can be drawn on the most critical components:

*Highest ranking*
First:
- spindle drive unit

Second equal:
- table traverse
- ball screws
- tool bit

| Equipment description | 1–3 ranking as impact on: | | | | | | | | |
|---|---|---|---|---|---|---|---|---|---|
| | M | R | PQ | ENV | TPV | KOE | S | C | Total |
| Supply panel (elec.) | 1 | 1 | 2 | 1 | 3 | 3 | 1 | 3 | 15 |
| Isolators | 2 | 1 | 1 | 1 | 3 | 3 | 3 | 2 | 16 |
| Motors | 2 | 1 | 1 | 1 | 3 | 3 | 2 | 3 | 16 |
| Control unit | 1 | 2 | 1 | 1 | 3 | 3 | 1 | 3 | 15 |
| Lighting | 1 | 2 | 1 | 2 | 1 | 1 | 2 | 1 | 11 |
| Cables | 1 | 1 | 1 | 2 | 3 | 3 | 2 | 2 | 15 |
| Pump (hyd.) | 2 | 2 | 1 | 3 | 2 | 2 | 1 | 3 | 16 |
| Pipework | 2 | 2 | 1 | 3 | 1 | 1 | 2 | 2 | 14 |
| Compressor (air) | 2 | 1 | 2 | 2 | 1 | 3 | 1 | 3 | 15 |
| Regulator | 2 | 1 | 2 | 2 | 1 | 1 | 1 | 1 | 11 |
| Lubricator | 2 | 1 | 1 | 1 | 2 | 2 | 1 | 1 | 11 |
| Moisture trap | 2 | 2 | 1 | 2 | 1 | 1 | 1 | 2 | 12 |
| Pipework | 1 | 1 | 1 | 2 | 1 | 2 | 2 | 1 | 11 |
| Pump (coolant) | 2 | 2 | 3 | 1 | 2 | 3 | 1 | 3 | 17 |
| Sump | 1 | 1 | 1 | 2 | 1 | 1 | 3 | 2 | 12 |
| Flexible nozzle | 1 | 1 | 2 | 2 | 2 | 3 | 3 | 1 | 15 |
| Pipework | 2 | 2 | 1 | 3 | 1 | 1 | 3 | 2 | 15 |
| Spindle unit (mech.) | 3 | 3 | 3 | 1 | 3 | 3 | 3 | 3 | 22 |
| Table traverse | 3 | 2 | 3 | 1 | 3 | 3 | 2 | 3 | 20 |
| Ball screw | 3 | 2 | 3 | 1 | 3 | 3 | 2 | 3 | 20 |
| Tow chuck | 1 | 1 | 3 | 1 | 2 | 2 | 2 | 2 | 14 |
| Worm chuck | 1 | 1 | 3 | 1 | 2 | 2 | 2 | 2 | 14 |
| Guards | 1 | 1 | 1 | 2 | 1 | 1 | 3 | 2 | 12 |
| Tool bit | 3 | 2 | 3 | 1 | 3 | 3 | 3 | 3 | 20 |

| | | 1 | 3 |
|---|---|---|---|
| M | = maintainability | easy | difficult |
| R | = reliability | high | low |
| PQ | = product quality | low | high |
| ENV | = environment | low | high |
| TPV | = throughput velocity | low | high |
| KOE | = knock-on effect | low | high |
| S | = safety | low | high |
| C | = cost | low | high |

**Figure 6.14**   *Miller M201 criticality assessment matrix*

*Lowest ranking*

First equal:

- lighting
- regulator
- lubricator
- coolant pipework

*Safety ranking*

First equal:

- coolant sump
- pipework
- flexible nozzle
- spindle drive
- tool bit
- guard
- isolators.

## 6.7 Condition appraisal

A completed condition appraisal form is shown as Figure 6.15.

## 6.8 Refurbishment programme

A study of the refurbishment possibilities indicated the following:

- tasks during machine operation    7    (20 hours)
- tasks requiring downtime    12    (34 hours)
- total tasks    19    (54 hours)

The costs of this programme are expected to be as follows:

- labour costs    £540
- material costs    £330
- total costs    £870

The major refurbishment tasks are:

- Renew slide blanket.
- Replace spindle gasket oil seal and bearing.
- Investigate vibration problem and cure.
- Clean and reorganize working area.

## 6.9 Asset care

A preferred spares listing for the M201 miller is shown in Figure 6.16.

Schedules for checking and monitoring and for daily cleaning and inspection are shown in Figures 6.17 and 6.18 respectively.

An operator training plan is drawn up in Figure 6.19.

---

CONDITION APPRAISAL - TOP SHEET

MACHINE No: __M201__        DESCRIPTION: INTERACT CNC

DATE INSTALLED: _____        MILLING MACHINE

COMMISSIONED: _____        _____

WARRANTY ENDS: _____        _____

LOCATION CODE: _____        MAKER: BRIDGEPORT

PLANT PRIORITY: _____        MANUFACTURER SERIAL No: _____

GENERAL GROUP: _____        EQUIPMENT STATUS: _____

P.O. NUMBER: _____        EQUIPMENT AVAILABILITY: _____

COMMON EQUIPMENT:

**GENERAL STATEMENT OF RELIABILITY**

GENERALLY REGARDED BY THE OPERATOR
AS A RELIABLE MACHINE. THE FOUR
MOST SIGNIFICANT RELIABILITY ISSUES ARE:-

1) ELECTRICAL FAILURE DUE TO SURGES/SPIKES.
2) VIBRATION PROBLEMS.
2) AIR SUPPLY FAILURES/LOW PRESSURE.
4) RUNNING OUT OF COOLANT (WHEN USED).

**GENERAL STATEMENT OF MAINTAINABILITY**

THERE IS NO PLANNED MAINTENANCE FOR
THIS MACHINE. THE OPERATOR DOES NOT
PARTICIPATE IN ANY MAINTENANCE ACTIVITY.
ACCESS FOR MAINTENANCE IS SEVERELY
HINDERED BY THE LAYOUT I.E. TABLES &
CUPBOARDS CLOSE TO THE MACHINE.

---

**Figure 6.15**   *Miller M201 condition appraisal record*

| | CONDITION APPRAISAL SHEET 1 OF | | | | |
|---|---|---|---|---|---|
| MACHINE DESCRIPTION: M 201, BRIDGEPORT C.N.C. MILLING MACHINE | | | | | |
| ASSET No: | | YEAR OF PURCHASE: | | APPRAISAL BY: | |
| MACHINE No: | | LOCATION: | | APPRAISAL DATE: | |

| ITEM No: | APPRAISAL RATING BY SUB ASSET | SATISFACTORY | BROKEN DOWN | NEEDS ATTENTION NOW | NEEDS ATTENTION LATER |
|---|---|---|---|---|---|
| | | X AS REQUIRED | | | |
| 1 | ELECTRICAL | | | | |
| | A - POWER SUPPLY TO MACHINE | | | X | |
| | B - PANEL | | | X | |
| | C - CONTROL | | | | X |
| | D - CONTROL CIRCUITS | | | X | |
| | E - MOTORS | X | | | |
| | F - MACHINE LIGHTING | | | X | |
| | G - OTHER | | | | |
| 2 | MECHANICAL | | | | |
| | A - SPINDLE HOUSINGS / GEARBOXES | | | X | |
| | - SEALS | | | X | |
| | - BEARINGS | X | | | |
| | - GEARS | X | | | |
| | B - SLIDEWAYS / TABLES | | | X | |
| | - WORKPLACE | | | X | |
| | - TOOLHOLDER | | | X | |
| | C - SCREWS / RAMS / SPLINED SHAFTS | | | X | |
| | D - PNEUMATICS | | | X | |

**Figure 6.15**   *(Continued)*

| | | SATISFACTORY | BROKEN DOWN | NEEDS ATTENTION NOW | NEEDS ATTENTION LATER |
|---|---|---|---|---|---|
| | CONDITION APPRAISAL SHEET 2 OF | | | | |
| 2 | MECHANICAL (CONTINUED) | | | | |
| | E - COOLANT SYSTEM | | | X | |
| | F - GUARDS | | | X | |
| | G - OTHER. | | | | |
| | | | | | |
| | | | | | |
| 3 | WORKING AREA | | | | |
| | A - LAYOUT | | | X | |
| | B - HAZARDS | | | X | |
| | C - STORAGE | | | | X |
| | | | | | |
| | | | | | |
| | | | | | |
| | | | | | |
| | | | | | |
| | | | | | |
| | | | | | |
| | | | | | |
| | | | | | |
| | | | | | |
| | | | | | |
| | | | | | |
| | | | | | |

**Figure 6.15**  *(Continued)*

| | CONDITION APPRAISAL SHEET 3 OF | |
|---|---|---|

ASSET No: M 201   DESCRIPTION: BRIDGEPORT MILLING M/c

LOCATION:

### SUB ASSET / GENERIC GROUP

DENOTE CONDITION AS ONE OF THE FOLLOWING:

**S - SATISFACTORY**
**B/D - BROKEN DOWN**
**NAN - NEEDS ATTENTION NOW**
**NAL - NEEDS ATTENTION LATER**

| GENERIC GROUP | PROBLEM FOUND | CONDITION |
|---|---|---|
| 1A | ELECTRICAL SYSTEM SUSCEPTIBLE TO SPIKES | |
| 1B | OPEN HOLE ON TOP SURFACE - SWARF/WATER | NAN |
| 1D | SYSTEM IS TEMPREMENTAL AT START-UP | NAN |
| 1F | CAUSES OPERATOR HEADACHES | NAN. |
| 1G | CABLES & PANELS COVERED IN SWARF | NAN. |
| | | |
| 2A | OIL LEAK ON SPINDLE HOUSING | |
| 2B | SLIDE PROTECTION BLANKET TORN/HOLED | NAN. |
| 2D | AIR SUPPLY PIPE TOO LONG— TRIP HAZARD | NAN |
| 2 E | KINKED PIPE | NAN |
| 2E | COOLANT TRAY NOT SECURE | |
| 2E | COOLANT TRAY DAMAGED. | |
| 2F | GUARD COMPONENTS NOT SECURE | NAN. |
| 2F | COMPLIANCE WITH CURRENT REGULATIONS? | NAN. |
| 2G | VIBRATION | NAN. |
| | | |
| | | |
| | | |
| | | |

**Figure 6.15** *(Continued)*

| | | CONDITION APPRAISAL SHEET 4 OF |
|---|---|---|

ASSET No: ............................................. DESCRIPTION: ...................................................

LOCATION: ...................................... ...........................................................

### SUB ASSET / GENERIC GROUP

DENOTE CONDITION AS ONE OF THE FOLLOWING:

S - SATISFACTORY
B/D - BROKEN DOWN
NAN - NEEDS ATTENTION NOW
NAL - NEEDS ATTENTION LATER

| | GENERIC GROUP | PROBLEM FOUND. | CONDITION |
|---|---|---|---|
| ✳ | 3A | CABINETS/TABLE/BINS RESTRICT ACCESS | N.AN. |
| ✳ | 3B | TABLE MOTION CREATES NIP POINT | N.AN. |
| | 3B | WET FLOOR - SLIP HAZARD (ROOF LEAK) | N.AN. |
| | 3B | SWARF EVERYWHERE INC. FLOOR. | N.A.N. |
| ✳ | 3C | NO DEFINED LOCATION FOR PARTS, INCOMING. OUTGOING OR SCRAP. | N.A.N. |
| | 3C | NO DEFINE LOCATION FOR BRUSH. HAMMER CHUCK KEY. SPARE TOOLS FILE | N.A.N. |
| | 3C | SETTERS CABINETS IN DISSARRAY | N.A.N. |
| | | | |
| | | | |
| | | | |
| | | | |
| | | | |
| | | | |
| | | | |
| | | | |
| | | | |
| | | | |

**Figure 6.15**  *(Continued)*

| PREFERRED SPARES LISTING | | | |
|---|---|---|---|

MACHINE No: __M201__   DESCRIPTION: __BRIDGEPORT__
LOCATION: __MACHINE SHOP__   __MILLING m/c.__

CRITICAL-■ DEDICATED-■ CONSUMABLE-■

| DESCRIPTION | QTY | PART No | SUPPLIER/TYPE |
|---|---|---|---|
| • GASKETS | 6 | | BRIGEPORT |
| • GAITORS | 6 | | " |
| • SLIDE COVERS. | 1 | | " |
| • SWARF BRUSHES | 12 | | LOCAL SUPPLY |
| • SAFETY GLASSES | 12 | | — " |
| • GLOVES | 12 | | " |
| • COATS | 12 | | " |
| • GUARD SPARES | 1 | | MANUFACTURERS |
| • COOLANT TRAY | 1 | | BRIGE PORT |
| • TOOL BITS (VARIOUS) | 20 | | LOCAL SUPPLY |
| • LUB - OIL            Lts | 200 | | " |
| • COOLANT (PARAFFIN)   Lts | 50 | | " |
| • COOLANT PIPES / CLIPS Mts | 10 | | " |
| • FUSES | 12 | | " |
| • BULDS | 12 | | " |
| • PAINT               Lts | 5 | | " |
| • LUBRICATOR | 6 | | " |
| • MOISTURE TRAP. | 6 | | " |
| | | | |
| | | | |
| | | | |
| | | | |
| | | | |

**Figure 6.16**   *Miller M201 asset care: spares listing*

| Frequency<br><br>1 = per shift<br>1a = per day<br>2 = per week<br>S = start<br>E = end | Shift | Oil reservoir/levels<br>S1 | Air lubricator inspect top-up<br>S1 | Moisture trap drain<br>S1 | Motor temperature<br>E1 | Pipelines air/oil/coolant<br>S1 | Vice security alignment<br>S1 | Table height<br>S1 | Warm-up cycle and emergency stops<br>S1a | Coolant levels<br>S1 | Air pressure<br>S1 | Vibration analysis motors and table<br>E2 | Electrical isolators<br>S1a | Maintenance dept. survey/inspections<br>E2 | Weekly clean routine oil slides and traverse gear<br>E2 | | Signature | |
|---|---|---|---|---|---|---|---|---|---|---|---|---|---|---|---|---|---|---|
| Monday | D<br>N | | | | | | | | | | | | | | | | | |
| Tuesday | D<br>N | | | | | | | | | | | | | | | | | |
| Wednesday | D<br>N | | | | | | | | | | | | | | | | | |
| Thursday | D<br>N | | | | | | | | | | | | | | | | | |
| Friday | D<br>N | | | | | | | | | | | | | | | | | |
| Saturday | D<br>N | | | | | | | | | | | | | | | | | |
| Inform maintenance | | | | | X | X | | | X | | X | X | X | | X | | | |

**Figure 6.17**  *Miller M201 checking and monitoring record*

| Frequency<br><br>1 = each component<br>2 = 2 per shift<br>3 = per shift<br>S = start of shift<br>E = end of shift | Shift | Clean M/C Remove swarf<br>2 | Empty swarf tray and clean filter<br>E3 | Brush and clean area around M/C<br>E3 | Clean M/C table<br>E3 | Clean M/C slides<br>E3 | Inspect and clean M/C head<br>E3 | Guards<br>E3 | Clean and inspect tool condition<br>S3 | Tool cleaning<br>1 | Vice and 3-jaw chuck<br>1 | Signature |
|---|---|---|---|---|---|---|---|---|---|---|---|---|
| Monday | D/S<br>N/S | | | | | | | | | | | |
| Tuesday | D/S<br>N/S | | | | | | | | | | | |
| Wednesday | D/S<br>N/S | | | | | | | | | | | |
| Thursday | D/S<br>N/S | | | | | | | | | | | |
| Friday | D/S<br>N/S | | | | | | | | | | | |
| Saturday | D/S<br>N/S | | | | | | | | | | | |
| Notify maintenance | | | | | | | X | X | | | X | |

**Figure 6.18**  *Miller M201 daily cleaning and inspection record*

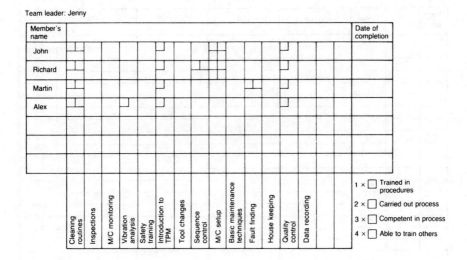

**Figure 6.19** *Miller M201 operator training plan*

## 6.10 Equipment history recording

Record forms for cycle time and downtime are shown in Figures 6.20 and 6.21 respectively.

| Week no. | | Day | Components | | | Changeover time | Stoppages | | Signature |
|---|---|---|---|---|---|---|---|---|---|
| | | | Produced | Rejects | Total | | Planned | Unplanned | |
| | D / N | Monday | | | | | | | |
| | D / N | Tuesday | | | | | | | |
| | D / N | Wednesday | | | | | | | |
| | D / N | Thursday | | | | | | | |
| | D / N | Friday | | | | | | | |
| | D / N | Monday | | | | | | | |
| | D / N | Tuesday | | | | | | | |
| | D / N | Wednesday | | | | | | | |
| | D / N | Thursday | | | | | | | |
| | D / N | Friday | | | | | | | |
| Totals | | | | | | (min) | (min) | (min) | |

**Figure 6.20** *Miller M201 equipment history record: operations*

| Inc No. | Date | Time | Equipment | Fault | Reason | Remedy |
|---------|------|------|-----------|-------|--------|--------|
|  |  |  |  |  |  |  |
|  |  |  |  |  |  |  |
|  |  |  |  |  |  |  |
|  |  |  |  |  |  |  |
|  |  |  |  |  |  |  |
|  |  |  |  |  |  |  |
|  |  |  |  |  |  |  |
|  |  |  |  |  |  |  |
|  |  |  |  |  |  |  |
|  |  |  |  |  |  |  |
|  |  |  |  |  |  |  |
|  |  |  |  |  |  |  |
|  |  |  |  |  |  |  |
|  |  |  |  |  |  |  |
|  |  |  |  |  |  |  |
|  |  |  |  |  |  |  |
|  |  |  |  |  |  |  |
|  |  |  |  |  |  |  |
|  |  |  |  |  |  |  |
|  |  |  |  |  |  |  |

**Figure 6.21**  *Miller M201 equipment history record: downtime and faults*

## 6.11  Assessment of overall equipment effectiveness

The OEE is given by the relation

$$OEE = availability \times performance \times quality$$

*Example OEE calculation*

The following example uses the data for the Tuesday of week 17 on the equipment history record of Figure 6.11.

Availablity

$$availability = \frac{uptime}{planned\ availability}$$

$$uptime = planned\ availability - downtime$$

$$downtime = unplanned\ stoppages + changeovers = 115 + 12 = 127\ min$$

Therefore

$$\text{uptime} = 910 - 127 = 783 \text{ min}$$

$$\text{availability} = \frac{783}{910} = 86.0\%$$

## Performance

$$\text{performance} = \frac{\text{completed cycles}}{\text{planned cycles}}$$

$$\text{completed cycles} = 203$$

$$\text{planned cycles} = \frac{\text{uptime}}{\text{standard cycle time}} = \frac{783}{3} = 261$$

Therefore

$$\text{performance} = \frac{203}{261} = 77.7\%$$

## Quality

(RFT = right first time)

$$\text{quality} = \frac{\text{components (RFT)}}{\text{completed cycles}}$$

$$\text{components (RFT)} = 152$$
$$\text{completed cycles} = 203$$

Therefore

$$\text{quality} = \frac{152}{203} = 74.9\%$$

$$\text{OEE} = 86.0\% \times 77.7\% \times 74.9\% = 50\%$$

## OEE

A summary of the OEE values for the equipment history provided (Figure 6.11) is given in Figure 6.22. A simple graph of the OEE possibilities is shown in Figure 6.23.

### Cost/benefit analysis

The cost/benefit analysis is based on the additional units that can be produced per week for each 1% improvement in OEE.

| Date | Availability | Performance | Quality | OEE |
|------|------|------|------|------|
| 15 M | 86.8 | 71.0 | 100* | 61.6 |
| T | 81.0 | 75.3 | 100 | 61.0 |
| W | 89.5 | 45.7 | 89.5 | 36.6 |
| Th | 83.4 | 82.6 | 100 | 68.9 |
| F | 84.5 | 68.3 | 92.6 | 53.4 |
| 16 M | 90.1 | 55.2 | 21.9 | 10.9 |
| T | 94.4 | 79.6 | 100 | 75.1 |
| W | 94.4 | 79.6 | 100 | 75.1 |
| Th | 86.9 | 45.9 | 100 | 39.9 |
| F | 94.5* | 70.1 | 100 | 66.2 |
| 17 M | 93.5 | 78.6 | 100 | 73.5 |
| T | 86.0 | 77.7 | 74.9 | 50.0 |
| W | 93.6 | 75.3 | 100 | 70.5 |
| Th | 66.0 | 56.6 | 100 | 37.4 |
| F | 93.1 | 83.6* | 67.8 | 52.8 |
| Average | 87.6% | 69.9% | 90.3% | 55.3% |
| *Best of best | 94.5% | 83.6% | 100% | 79.0% |

Difference between best of best and average: 24%

**Figure 6.22**    *Miller M201 OEE summary*

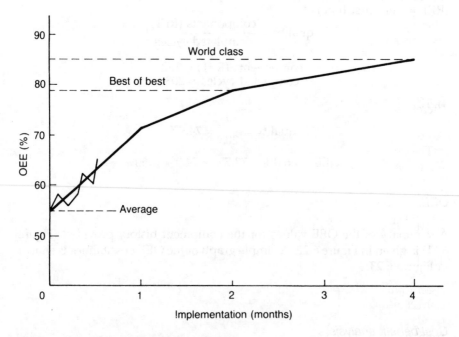

**Figure 6.23**    *Miller M201 OEE comparison*

$$\text{units/week/\%OEE} = \frac{\text{total components (RFT)}}{\text{average \%OEE} \times 3 \text{ weeks}}$$

$$= \frac{2546}{55 \times 3} = 15.43$$

Thus the benefit of increasing the average OEE up to the best of best OEE (approximately $79 - 55 = 24\%$) is equivalent to an extra 370 units per week.

## 6.12 Assessment of the six losses

The problems identified for the M201 are as follows:

- abnormal operation
- electrical power loss
- vibration
- no air supply
- tooling performance affected
- long cycle time
- initial startup procedure
- excessive component loading time
- slideway damage
- operational safety
- no reference documentation.

These problems have been allocated to the six losses in Figure 6.24. Part of the loss assessment record is shown in Figure 6.25.

## 6.13 Problem solving and improvements

An example of a problem solving document is shown in Figure 6.26.

The improvements identified, and their effects on availability, performance and quality, are shown in Figure 6.27.

## 6.14 Best practice routines

The key areas for attention to best practice are as follows:

*Asset Care*   Cleaning; monitoring; planned maintenance.
*Correct operation*   Clear instructions; easy to operate; understand process.
*Good support*   Maintenance and operator work together. Additional support from accounts, production, design, purchase and planning (the key contacts).
*Inspection*   Operator's responsibility.
*Training*   Operator and maintenance.

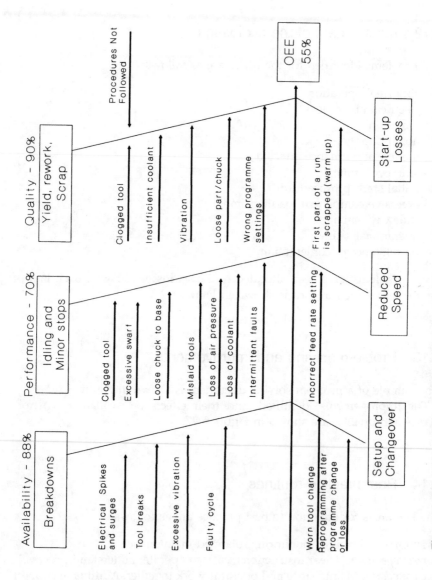

**Figure 6.24** Miller M201: problems identified in the assessment of the six losses

| Loss type | Item | Availability | Performance | Quality | Associated problems issues | Impacts |
|---|---|---|---|---|---|---|
| Breakdown | Electrical surge/spike | ◉ | | ◉ | • Lost/corrupted program requires setter to reload and test<br>• Scrapped part<br>• Possible damage to equipment | • Immediate stop<br>• Additional startup cycle<br>• Program and inspection tests |
| Breakdown | Tool break | • | • | ◉ | • Scrapped part<br>• New tool to be fitted<br>• Setter needed to restart | • Immediate stop<br>• Purchase new tool |
| Idling and minor stops | Clogged tool | | • | ◉ | • Affects surface quality<br>• Redues cutting efficiency, hence increases load on machine, accelerates wear<br>• Operator needs to clean tool | • Reduced performance<br>• Progressive buildup affecting quality, not identified by operator: scrap parts |
| Idling and minor stops | Loose chuck | | • | ◉ | • Affects part quality (surface and size)<br>• Operator needed to retighten | • Reduced performance<br>• Progressive loosening affects quality, not identified by operator: scrap parts |
| Yield/rework | Insufficient coolant | ◉ | | • | • Affects part dimension (growth)<br>• Reduces tool life<br>• Increases load on machine | • Reduced quality<br>• Reduces availability due to tool change |

• = primary impact
◉ = secondary impact

**Figure 6.25** Miller M201 loss assessment record

---

## TPM PROBLEM SOLVING DOCUMENT

MACHINE / PART NAME: M201 MILLER          TEAM LEADER: ............................

PROBLEM RAISED BY: ...........................          DATE: 7/9/93

---

**1. PROBLEM STATEMENT (SPECIFIC)**

Excessive Cycle Time resulting in low output

---

**2. CLARIFICATION OF PROBLEM (BY WHOM, WHERE, WHEN AND HOW) PLUS COST / BENEFIT OPPORTUNITY**

Clarified with operator on 7/9/93.

Discussed opportunities to reduce cycle time without sacrificing product quality.

Cost to modify programme ~ 1 hour set up + 1 hour test/inspect.

Benefit. Standard cycle time reduced from 3 to 2 seconds = extra 152 units per week (~ 10% OEE improvement).

---

**3. PROBLEM CAUSE (BRAINSTORM AND FISHBONE, THEN LIST)** 8/8/93.

CAUSED BY 1: Shallow depth of cut

CAUSED BY 2: Movement of table in z-plane

CAUSED BY 3: Relative position of tooling start position and workpiece

CAUSED BY 4: Need to manually reset programme to start cycle due to multiple programmes in memory

CAUSED BY 5:

---

**Figure 6.26**   *Miler M201 problem solving document*

# TPM PROBLEM SOLVING DOCUMENT

**4. CLARIFICATION OF ROOT CAUSES (DO THE CAUSES EXPLAIN THE PROBLEM?)**

- Single 10mm cut is practical rather than 5 x 2mm cuts.
- If single cut implemented table height can be fixed via programme
- Reduction of tool travel shortens cycle time.
- Programme location/multiple programs prevent auto reset function being utilised thus manually reset for each cycle.

**5. COUNTER-MEASURE (ACTIONS REQUIRED TO RESOLVE CAUSE(S))**

**TEMPORARY:**

—

**PERMANENT:**

- Modify programme to single cutting cycle from 5 cuts.
- Set traverse table at constant height in programme.
- Reprogramme tool start position closer to workpiece
- Store single programme in memory and locate at line 0 to enable auto-reset statement to be used.

**6. CONFIRMATION OF COUNTER-MEASURE (HAVE ACTIONS CLEARED PROBLEM?)**

COUNTER-MEASURES AWAITING IMPLEMENTATION

**7. FEEDBACK (WHO ELSE NEEDS TO KNOW?)**

MILLER SUPERVISOR, MAINTENANCE MANAGER; TPM FACILITATOR,
NOMINATE TEAM FOR TPM EXCELLENCE AWARD (SEPT 93).

**Figure 6.26**   *(continued)*

| | | Benefit for: | |
|---|---|---|---|
| | *availability* | *performance* | *quality* |
| Reduce cycle time | | high | |
| Program modficiations | | high | |
| Operator to load programs | low | | |
| Use of coolant | low | | medium |
| Operator to change tooling | medium | | |
| Reposition air supply service unit | low | | |
| Define startup procedure | | high | low |
| New slideway cover design | low | | |
| Obtain copies of manuals/drawings | low | | |
| Replace with new design guards | low | | |
| Operator training programme | high | high | high |

**Figure 6.27**   *Miller M201 improvements*

| | Sept. | Oct. | Nov. | Dec. | Jan. | Feb. | Mar. | Apr. | May | June | July | Aug. | Sept. |
|---|---|---|---|---|---|---|---|---|---|---|---|---|---|
| Refurbishment | | | | | | | | | | | | | |
| During M/C oper. | | | | | | | | | | | | | |
| M/C shut | | | | | | | | | | | | | |
| Improvements | | | | | | | | | | | | | |
| Asset care | | | | | | | | | | | | | |
| Cleaning | | | | | | | | | | | | | |
| Cond. mon./planned main. | | | | | | | | | | | | | |
| Review OEE | | | | | | | | | | | | | |
| Best practice | | | | | | | | | | | | | |
| Stores review | | | | | | | | | | | | | |
| Training | | | | | | | | | | | | | |
| Maintenance | | | | | | | | | | | | | |
| Operators | | | | | | | | | | | | | |

**Figure 6.28**   *Miller M201 implementation programme*

## 6.15 Implementation

The proposed implementation programme is shown in Figure 6.28.

## 6.16 Conclusions

To achieve world class performance, the company must move forward to

- achieve a total quality process
- satisfy customer requirements
- consider all internal and external factors
- meet internal customers and suppliers and agree requirements.

TPM is the vehicle to deliver customer satisfaction and to secure the company's future and jobs. In this pilot project, the meaning of TPM may be said to have moved from 'today's problematic miller' to 'tomorrow's perfect miller'!

Suffice to say that this syndicate team won the award of 'best presentation' adjudged by their other three syndicate team colleagues on the final day of their four-day TPM facilitators' training course.

# ——— 7 ———

# *Planning and Launching TPM*

The improvement plan described in Chapters 5 and 6 is concerned with the means to achieve continuous improvement in overall equipment effectiveness. This chapter explains how the scope of the TPM process is determined and goes on to examine the provision of resources, the approach to implementation, the pilot project(s) as the forerunner to plant-wide application, the training programme, and the operation and control of the project.

Before any of these processes can take effect, the company management must examine the existing situation under seven fundamental headings:

- people's perceptions and feelings
- the present organization and structure
- the commitment of the company to change
- the likely pace at which change can take effect
- the status of other change initiatives (e.g. total quality)
- the condition and type of plant and equipment
- the processes used in the plant.

Study and assessment of these seven heads will disclose the extent to which they may assist or inhibit the introduction of TPM. The study will also develop a measurement strategy, an assessment of potential benefit, the priorities, and the integration of the elements of the plan. The scope, implementation and training plans can then be tailored in accordance with the findings of the study.

The headings in this chapter indicate the approximate order in which the phases of the TPM plan take place, but these phases are not absolute and there are overlaps. Some new phrases in the language of TPM will appear and these are explained as appropriate. Figure 7.1 shows the TPM implementation route, as introduced in Chapter 3.

## 7.1 TPM assessment

The assessment takes the form of in-depth interviews and discussions to ascertain existing attitudes and in due course to influence those attitudes.

In WCS we have developed a 20 statement format which employees (whether they are the managing directors, design engineers, operators or maintainers) are asked to rank the statements from their own perspective and perception. For example the statement might say:

'From my viewpoint, Production and Maintenance operate as separate empires'. Do you think this statement is very true, partially true or false?

Ten of the statements measure the employee's perception with regard to the degree of management encouragement in the organization or plant, and the

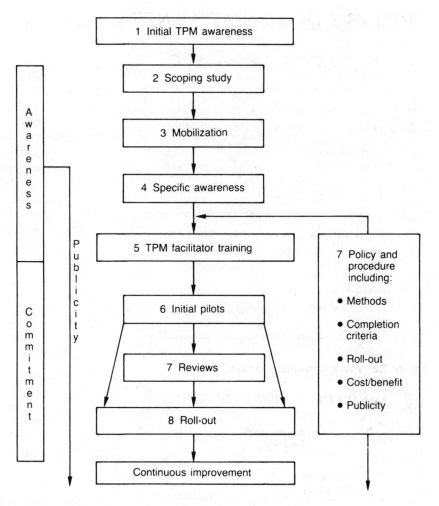

**Figure 7.1** *TPM implementation route*

other ten statements measure the degree of workforce involvement (see Figures 7.2 and 7.3).

It is far better to carry out these perception interviews on a one-to-one basis rather than simply giving out the questionnaire to be completed by the employee since:

- it gives the interviewer a chance to explain TPM and how and when it might affect the employee
- it allows the interviewer the opportunity to ask a supplementary question to each of the 20 statements such as 'Why do you feel so strongly about this statement?' The response will often give some key directions and insights as well as, perhaps, an improving or occasionally worsening perception over time.

# *ANALYSIS OF 20 STATEMENTS*

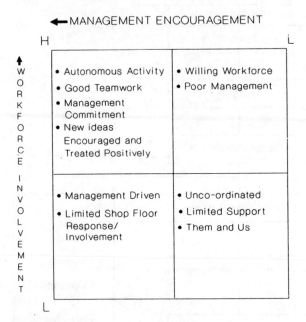

**Figure 7.2**   *Perceptions matrix analysis*

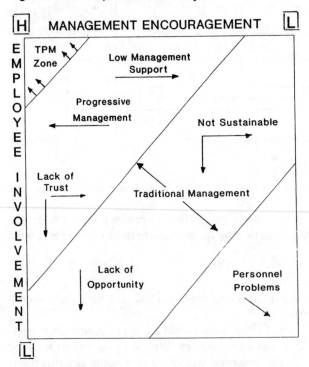

**Figure 7.3**   *Perceptions matrix implications*

The subsequent analysis which clearly shows the differences in strengths of feeling between, say, members of the management compared to key contacts and, of course, comparisons with operators and maintainers. Similarly, the strength of response across the spectrum of employees will show up quite clearly for each of the twenty statements. These can then be grouped as 'hopes' and 'fears' as well as potential TPM 'hinders' and 'helps'.

The analysis will position employee groups on the matrix; the higher the grouping towards the top left hand corner, the better for TPM's likely acceptance and success. However, if the groupings are towards the lower end of the horizontal and vertical axis, you will find that the TPM process addresses in a positive and lasting way, many of the perceived hindrances. This perceptions tool is not absolute, but it does provide an excellent benchmark against which to measure future movements on the matrix.

The analysis will have a major bearing on the way the TPM process is implemented. As the plan develops the training programme will seek to ensure that the most constructive and progressive attitudes prevail, firstly in the pilot project (see later) and then company-wide as the TPM process develops.

Achieving the right attitude to change is essential for success. Experience has shown that operators, recently engaged staff and younger people tend to take a positive attitude to change, whereas the old hands and the experienced maintenance technicians are likely to be more wary and defensive (Figure 7.4). The attitude of supervisors depends very much on the individual. Supervisors will normally support the idea of TPM because of its common sense. However they have to face the day-by-day demands of production, quality and research and hence may find it difficult to sustain a commitment to release operators and maintainers for the TPM process or to release equipment and machines for essential restoration and refurbishment. Effective two-way communication is essential to avoid resistance to change: those who will be involved in the TPM process must have a very clear idea of what it is all about and what the company – and, more particularly, what they as individuals – stand to gain. Resistance must be broken down by explanation, thorough discussion and the establishment of total confidence in the eventual outcome (Figure 7.5).

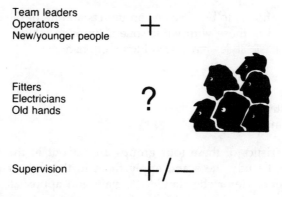

Team leaders
Operators
New/younger people $+$

Fitters
Electricians
Old hands $?$

Supervision $+/-$

**Figure 7.4** *How they see TPM*

---

Poor communications = resistance

- The less I know about plans to change, the more I assume;
- the more I assume, the more suspicious I become;
- And the more I direct my energy into

**RESISTANCE**

---

**Figure 7.5**   *Causes of resistance*

Figure 7.6 depicts the way in which resistance can be broken down by ensuring the full involvement of the people concerned and by securing their enthusiasm and dedication. Effective communication is more than the one-way approach of informing people and preparing them for change. People need to be part of the change process, actively involved in decisions and able to influence the outcome. If treated with respect they gain recognition and self-esteem and a two-way communication will result. They may well need to get rid of the acid or bile, which has built up over years, out of their system, but essentially employees *want* to improve their lot. A typical reaction came from a team leader, who, after three months' involvement in the TPM process, commented that TPM was all about Teamwork between Production and Maintenance and that 'Today People Matter'.

If Jack is said to have an attitude problem it is normally assumed that he has a *negative* attitude. He has every right to feel that way and may well have been influenced by developments over a number of years. It will take time to change negative energy into positive energy, and TPM may well be the catalyst to move Jack from left to right on Figure 7.7.

## 7.2   Operating style

If communication is to be effective, an understanding of the operating style of oneself and of those with whom one is working is well worthy of close study. Four basic styles have been identified, namely:

- delegator
- autocrat
- administrator
- player/manager.

The characteristics of these four groups are set out in the following. The reader should study these and apply them to the task of ensuring good communication. He or she should ascertain and appreciate *his or her own style* as well as that of others (Figure 7.8).

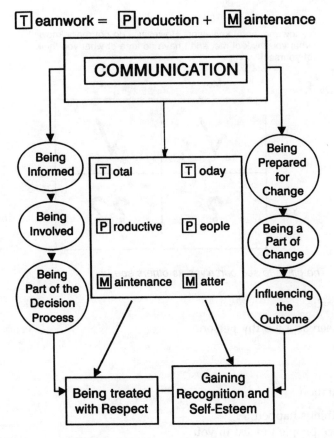

Ｔeamwork = Ｐroduction + Ｍaintenance

**Figure 7.6**   *Breaking down resistance*

| Phase | | Negative | Positive |
|---|---|---|---|
| 1(a) | Awareness | Not interested | Seeking out new ideas |
| 1(b) | Commitment and preparation | Cynical: it won't work | Active learning |
| 2 | Transfer to real life and refine | It's flavour of the month | Objective testing |
| 3 | Build capability and take off | Highlighting failures | Making it work |
| 4 | Continuous improvement | It's only what we have always done | Refining, improving |

**Figure 7.7**   *Attitude to change: indicators of progress towards a TPM culture*

*Basic profiles of operating styles*

Delegator

Basic orientation:

- responsive to people
- pursue excellence and ideals
- do a good job and good things will come to you.

I know me, and I know you. However, I do not really know what you think of me, and I have no idea of what you think of yourself!

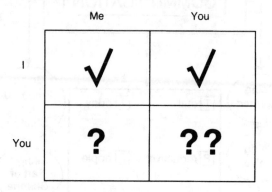

**Figure 7.8**   *The gift is to see ourselves as others see us*

Personal goal:

- to be seen as a worthy person.

## Autocrat

Basic orientation:

- make things happen
- convince people to trust in you
- you cannot wait for things to come to you.

Personal goal:

- to be seen as competent.

## Administrator

Basic orientation:

- preserve what you have
- use existing resources
- build the future on the past.

Personal goal:

- to be seen as objective and rational.

## Player/manager

Basic orientation:

- be sensitive to the needs of others
- adapt yourself to others, filling their needs first.

Personal goal:

- to be seen as likeable and popular.

## Strengths and weaknesses

### Delegator

Strengths:    cooperative, trusting.
Weaknesses:   easily influenced, gullible.

### Autocrat

Strengths:    quick to act, self-confident.
Weaknesses:   impulsive, arrogant.

### Administrator

Strengths:    economical, methodical.
Weaknesses:   stingy, plodding.

### Player/manager

Strengths:    flexible, adaptable.
Weaknesses:   inconsistent, without conviction.

## How to communicate with each style

### Delegator

- Stress worthy causes.
- Ask for help.
- Show concern.
- Emphasize self-development.

### Autocrat

- Offer opportunity.
- Give responsibility.
- Challenge.
- Give authority.

### Administrator

- Present ideas as low risk.
- Exercise logic.
- Tie new things to old.
- Use familiarity, routine and structure.

### Player/manager

- Chance to do things with others.
- Use humorous appeals.
- Let them know you are pleased.
- Provide opportunities to be in the spotlight.

## *Time management styles*

### Delegator

| | |
|---|---|
| Productive use of time: | seeking excellence. |
| Time trap: | saying yes to too many people. |
| Time concern: | missing a chance to be appreciated. |
| Key phrase: | 'Here's what I need.' |

### Autocrat

| | |
|---|---|
| Productive use of time: | generating new approaches. |
| Time trap: | trying to do too much. |
| Time concern: | losing an opportunity. |
| Key phrase: | 'Here's the thing for today.' |

### Administrator

| | |
|---|---|
| Productive use of time: | analysing and getting facts. |
| Time trap: | analysis/paralysis. |
| Time concern: | not understanding the situation. |
| Key phrase: | 'Here's what I'm going to do.' |

### Player/manager

| | |
|---|---|
| Productive use of time: | resolving conflicts. |
| Time trap: | always open to new change. |
| Time concern: | loss of approval. |
| Key phrase: | 'Here's what I think should be done.' |

*Team styles*

Each style can be useful in team situations.

## Delegator

- Provides help to others within own organization.
- Sensitive to environmental pressures.

## Autocrat

- Organizing people and resources.
- Directing tasks.

## Administrator

- Planning projects.
- Controlling tasks.
- Problem solving.

## Player/manager

- Negotiated settlements.
- Client-oriented situations.

*Style reactions to change*

### Factors that facilitate change

| | |
|---|---|
| Delegator: | involvement in group planning. |
| Autocrat: | dynamic presentation. |
| Administrator: | practical and cost payoffs. |
| Player/manager: | reputation building. |

### Management patterns

| | |
|---|---|
| Delegator: | likes to delegate. |
| Autocrat: | keeps power to self. |
| Administrator: | formal. |
| Player/manager: | concerned about reactions. |

### First reaction

| | |
|---|---|
| Delegator: | hesitant, must see the benefit. |
| Autocrat: | favourable, if it supports self-interest. |
| Administrator: | cautious, wait and see. |
| Player/manager: | willing to have a go. |

### Strength for coping with change

| | |
|---|---|
| Delegator: | loyalty, goal orientation. |
| Autocrat: | initiative, energy. |
| Administrator: | analysis, objectivity. |
| Player/manager: | reads people well, bolsters spirits. |

### Key factors for accepting change

| | |
|---|---|
| Delegator: | request ties with values. |
| Autocrat: | can stay in control. |
| Administrator: | proven, sound. |
| Player/manager: | reputation. |

### When change is difficult

| | |
|---|---|
| Delegator: | tries, but gets discouraged. |
| Autocrat: | will persist, but loses interest if nothing happens soon. |
| Administrator: | stops, assesses difficulty, slows down. |
| Player/manager: | gets concerned, sees if others can help. |

## 7.3   Scoping study

The scoping study is the second stage of the TPM implementation route (Figure 7.1). It   will determine the way in which the TPM process will develop, and is likely to start with an assessment of the seven fundamentals with which this chapter opened. The study will be concerned with a number of other aspects, all of which will be covered in this chapter:

- benefits likely to result from reduction of losses and improvements in equipment effectiveness
- training needs for successful implementation of TPM
- areas of the company's activities which are seen to be likely candidates as pilot projects
- makeup of teams and squads for the pilot projects
- availability of the resources necessary for implementation of the pilot project and later extensions to plant-wide activity
- scope of initial awareness workshops and later the training plan
- support from facilitators and from the TPM steering group
- initial roll-out plan for the whole site.

Figure 7.9 shows a typical outline programme for a scoping study, and Figure 7.10 shows progress towards the selected pilot project. The following is taken from briefing notes for a scoping study team.

## Initial awareness

The initial awareness session represents an opportunity to pick up any people who were unable to attend earlier sessions.

## Site tour

It is necessary to tour the site for about two hours with a knowledgeable person.

- Agree potential pilot areas and equipment.
- Find and photograph key problem machines and process/maintenance issues, including CAN-DO site, maintenance tools and information

plus 2 squad support
Total 20 people to see, or
as representative
% of total

Up to 20 one-hour interviews

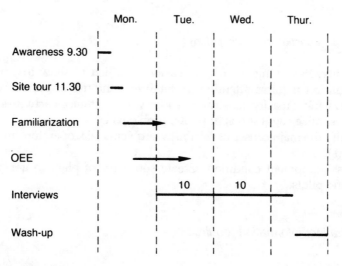

**Figure 7.9** *Scoping study outline timetable*

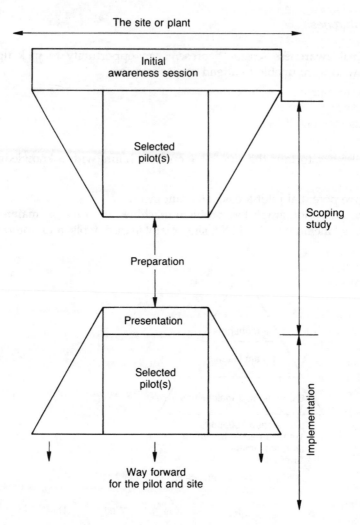

**Figure 7.10**   *Function of scoping study*

visibility. Pick things which are easily identified to show that problem
recognition requires attention to detail and the use of our senses.
- Understand capacity constraints and review bottleneck activities. In the
  groups, raise suggestions of pilots, problems etc.
- Identify the main causes of reliability problems (ask operators and maintainers).
- Assess equipment condition. Create summary of plant in general and
  specific pilots.

*Familiarization: information required*

- Site plan/map
- Total annual value of sales transfer from the plant.

- Total employees on the site:
    management
    indirect support
    direct.
- Senior management organization structure for the plant.
- Current process organization structure for the plant and specifically for the potential pilots, showing:
    management
    supervision
    hourly paid skilled and semi/unskilled by job title, by process location, by shift pattern
    zoning and central maintenance and organization.
- Maintenance revenue budget spend and actual:
    manufacturing process
    utilities
    buildings and infrastructure
    value of spares purchases
    stock valuation of spares
    capital spend.
- Amount and reasons for subcontract.
- Amount and reason for overtime.
- Flow chart of the main processes, especially the potential pilot areas.
- Main/critical plant items/path in the process regarding the potential pilot areas.
- Working arrangements: shift patterns, shutdowns, product changeovers.
- Daily/weekly/monthly output budget/targets/actuals for manufacturing process.
- Typical log sheets and fault analysis data.
- How is the plant performance, quality, availability recorded and managed? Weekly/daily meetings? Any minutes produced?
- Maintenance work scheduling, stores location, stock control.
- Balance between predictive condition based, planned preventive, and breakdown.
- Strategic policy statements and pictures of intentions over the next few years, especially those relating to total quality, world class visions etc.

## OEE data collection

- Get company to produce data: usually have good data on bottlenecks. Suggest four-week reference period.
- Document process flow, manning positions and stocking points.
- Document shift pattern using a 24-hour, 7-day format to show shift overlaps for maintenance, operators and management.
- Identify level of training in areas such as teamworking, problem solving, employee involvement, participative management.
- Identify the last improvement project(s) and their level of success, both real and perceived.

- Analyse responses to the 20 statements. This will support the other two analyses described below and will identify the perceived reality at the plant.

### Interviews

Minimum of twenty: say six maintenance, six operators, six support and two potential TPM facilitators, or a percentage of the workforce.

- Sell TPM: 15 minutes using handout and visual aids.
- Understand perceptions.
- Identify problem areas.
- Obtain quotable quotes.
- Identify potential pilots, team members and facilitators.

### Wash-up meeting

- Agree pilot areas with approximate timescales including a potential roll-out programme.
- Agree TPM core team and key contact profile, facilitator, project champion, steering committee format.
- Identify how TPM pilot team meetings can be held within the current shift patterns.
- Agree how TPM fits into company goals. (Why are they doing this? How will TPM help them to stay in business?)
- Confirm training programme focus (if possible, provisional start dates).

### Analysis of data

1   Assessment of the company against the ten TPM win commandments (see Figure 3.12). This is used to establish the risk of failure and therefore the level of support needed for a successful implementation.
2   Assessment of the scoping study data against the profile of ten key variables. This is used to determine the inhibitors which will need to be overcome during the pilot exercise. The ten variables are:
    - people's perceptions and feelings
    - organization structure
    - commitment to change
    - pace of change
    - status of change
    - equipment condition, type
    - processes and types
    - measurement
    - potential benefit
    - priority and integration.

## Scoping study output

This is in the form of a presentation to the management and key contacts. The presentation is supported by bound copies of the visual aids in the following format:

- Introduction: aims, how we got to this stage etc.
- TPM concepts: nature of TPM, how it supports the business vision.
- Scoping study results: potential financial benefits, perceptions and attitudes, effectiveness inhibitors, quotable quotes.
- Pilot strategy: including team, key contacts etc. and proposed pilots.
- Programme timetable: including awareness and training modules and key milestones.
- Conclusions and the way forward.

## Other points

Where possible combine the scoping study with discussion on mobilization, so that the emphasis of the feedback meeting is on confirming start dates and who should be involved.

## 7.4  Facilitators

Facilitators will need to be trained in the methods of TPM, and it will be their function to help all those involved in the process, to explain the methods and to coordinate the effort. Their role will be to influence rather than to do, to be effective without direct authority, and to get the teams to deliver results and to develop their capability.

Figure 7.11 indicates an example from an automotive plant of the structure necessary to provide facilitators on a plant-wide basis. A typical job description follows.

## Job purpose

To establish total productive maintenance practices across the company/ plant/unit so that it can achieve world class levels of equipment effectiveness based on a principle of continuous improvement.

These practices are characterized by the focus on reducing the causes of accelerated deterioration, improving the quality of maintenance and passing on the lessons learned from problem solving to achieve the maximum benefit.

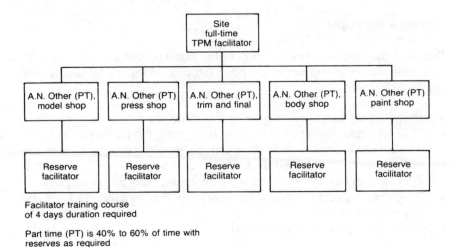

Facilitator training course
of 4 days duration required

Part time (PT) is 40% to 60% of time with
reserves as required

**Figure 7.11** *Typical plant-level facilitator support*

## Reporting

This is a support function which should report directly to senior management and/or the main board TPM champion. This will make it easier to influence opinion.

## Principal accountabilities

- To support the development of a common vision, across all functions and departments, integrating TPM philosophies and practices within their daily activities.
- To establish systems and procedures for setting and monitoring equipment effectiveness goals at a local and plant-wide level.
- To define, together with the appropriate managers, clear responsibilities for the reliable completion of TPM activities.
- To agree local targets and strategies for continuous improvement in equipment effectiveness, ensuring ownership at all levels of the organization.
- To identify the level of resources to support TPM activities and influence management so that they are made available.
- To ensure that momentum is maintained between TPM activity days so that the time spent by the teams can be as productive as possible. This might include establishing links with key contacts as required, briefings, data collection, photographs, typing work or form development.
- To promote the development of individual capability to identify and solve problems and progressively move towards the vision of the future.
- To reinforce the essential cross-departmental teamwork through the use of communication or activities to highlight the benefits of the approach.

- To campaign for a positive attitude towards TPM so that individuals will see the benefit and support the process with a positive attitude.

## Nature and scope

The nature of the facilitator's role is to influence rather than do. However, the facilitator has a responsibility to build ownership and capability so that individuals *want* to carry out TPM activities.

The TPM facilitator should be a central point for information on TPM and encourage individuals to understand and use TPM concepts. The facilitator should also ensure that TPM activities are coordinated with other related initiatives, taking every opportunity to increase awareness of the benefits of TPM.

## Measuring success

Success will be measured by the achievement of a self-sustaining process.

Ongoing improvement in the OEE provides a quantitative measure of this success. It is important, therefore, that this is developed in such a way that it can be applied consistently and is easy to use. Ideally target improvements in OEE should be linked to the three-year to five-year business goals of the company.

## Managing the task

The TPM facilitator will need to influence opinion at all levels to achieve this without direct authority.

The effectiveness of the role depends on a number of factors. This includes experience, interpersonal skills, attitude and the amount of time which can be dedicated to the task.

Ideally the main facilitator should be a full-time role. During the roll-out programme, supporting facilitators may be appointed on a part-time basis but will still need to allocate around 60% of their time to TPM activities.

This requires careful management of time. The combination of the facilitator role with other activities is difficult where the other role includes line responsibilities with the need for daily input. It is much more realistic to combine the role of the part-time facilitator with indirect or other project responsibilities.

## Person specification

- Preferably some maintenance experience, although broadly based production experience is perhaps more important.

- Ability to communicate concepts, principles, ideas and new ways of working. Therefore real desire to plan and implement effective change and an ability to teach, convince and relate to the practicalities of the specific plant environment.
- Age range 30 to 45 years: could, however, be 50+ if he/she has the necessary energy and experience. Unlikely to have necessary credibility/experience if under 25 years.

## 7.5   Teams and key contacts

The concept of teams and key contacts was introduced in Chapter 3 and an analogy with a successful soccer team was made in that chapter. The team exists to make things happen on the production line by progressive implementation of the TPM process.

The characteristics of a high-performance team are:

- mutual trust and support
- good communication
- shared objectives
- managed conflict
- effective use of skills.

Managed conflict involves allowing different points of view and different objectives to be aired and resolving them to the mutal benefit of the team.

Some of the negative practices which must be eliminated are:

- over-talking
- not joining in
- going off at a tangent
- hijacking the discussion.

All of these can be avoided by good team leadership. Good decisions will be reached if the leader

- searches for the common ground
- elicits support for or opposition to the consensus
- summarizes and ensures full agreement.

The makeup of the key contacts and their relationships with the teams are shown in Figure 7.12. The means by which key contacts can help teams to succeed are:

- agreed priorities and strategy
- effective planning, control and delivery systems
- clear organization of labour, equipment and materials
- insistence on measurable results and individual commitment to them
- encouragement to identify and meet task and process skills needs
- active reinforcement of teamworking
- promotion of a positive outlook to problem solving and new ideas.

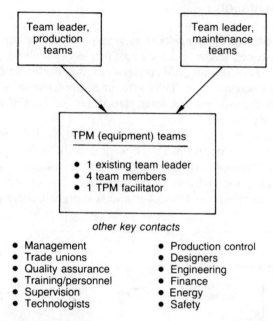

**Figure 7.12** *Typical TPM team*

*Key contact support, membership and roles*

The key contacts will be:

- business manager
- engineering manager
- TPM facilitator.

The roles of the key contacts are to provide direct support to the team when and if required, and to help improve communications between departments to achieve a consistent quality of TPM implementation. They are also to provide sustained direction and support to the team, to achieve continuity and to ensure *communication between shifts*. Further key contact tasks are as follows:

- technical and historical information
- plant and line OEEs
- commercial and market benefits definition
- visual management of information
- TPM publicity and awareness communication
- TPM activity logistics and facilitator support
- standardization of best practice
- spares forecast and consumption rates
- hygiene and/or safety training and policy
- input to problem solving and solutions support.

## 7.6   Implementation

Figure 7.13 illustrates the long-term aspects of TPM implementation, and shows how the four phases of the change process (see also Figure 7.7) and the key milestones of training, awareness, scoping studies and initial pilots fit into the total scheme. The TPM implementation route is portrayed in Figure 7.1, and this shows the interrelationship of all the factors which contribute to the plan.

Implementation of TPM encounters the 'bow wave' effect which is explained in detail in Chapter 8. The concept is really quite simple: preparation, training, awareness and time devoted to discussion and evaluation all have a cost, and this cost must be provided for. The upshot is that when looking at cost/benefit from TPM over a long timescale there will be, in the

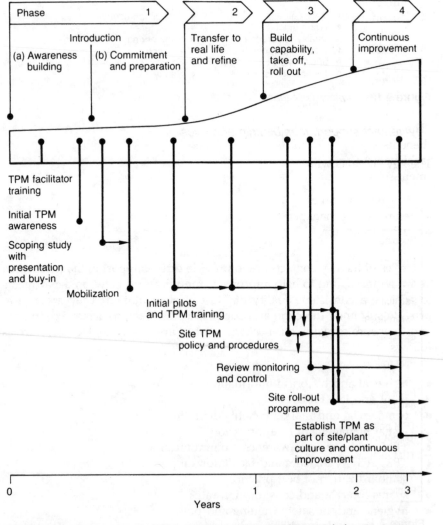

**Figure 7.13**   *TPM implementation milestones: improvement and change process*

early stages, a 'negative' benefit, that is a cost. As implementation proceeds and the real long-term savings are achieved, this initial surge in costs will be covered many times over by the benefits.

A typical implementation plan covering the first year is illustrated in Figure 7.14. The result of the inital pilots, when presented, will cover:

1 Milestones and detailed (three-year) forward programme.
2 Condition appraisal method and progress including:
  • refurbishment programme
  • spares and labour requirements
  • future planned, preventive maintenance schedules
  • daily, weekly, monthly cleaning, inspection and check schedules with clearly identified responsibilities as to who does what.
3 Equivalent OEE measurement method and cost/benefit appraisal targets linked to timescales.
4 Standard practice operating and asset care instructions (i.e. best practice routines) with the use of sketches and photographs.
5 Visibility controls and workplace organization with the use of simple graphics, before and after photographs, plus failure records.
6 Potential improvement areas and problem solving as an assessment of the relevant six losses and the potential for reduction/elimination.

The above approach is the practical acid test for showing the real on-the-job benefits of TPM. The true goal is to get ownership, pride and a sense of achievement through teamwork on the shop floor, and to eliminate equipment-related problems *once and for all*.

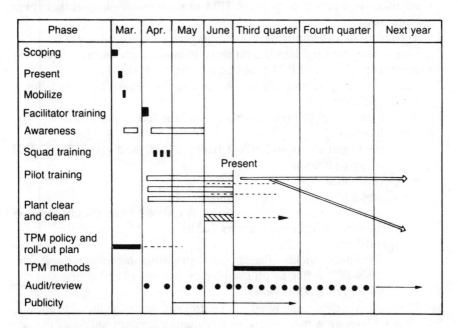

**Figure 7.14**  *Site roll-out plan*

## 7.7   Pilot project

One of the many outputs from the scoping study will be proposals for the pilot projects which will, in effect, try out the TPM process in a selected area of activity. The area selected should enable eventual development of a plant-wide scheme and must, therefore, be chosen with this development in mind. Vital lessons will be learned from the pilot project and these can be applied as the process of implementation proceeds.

The objectives of the pilot project may be summarized as follows:

- act as a lever to change attitudes
- demonstrate the power of TPM
- prove the nine-step, three-cycle TPM improvement plan
- show effective teamworking
- objectively test the water
- address hindrances
- build a TPM policy and procedures bible
- plan for roll-out and infrastructure.

Most important of all, the project will provide direct experience for those involved, enable others to observe the way the process works, and gradually spread the TPM message to other areas.

The pilot project will focus on a specific machine, equipment or process, and the work will be supported by on-the-job coaching which will cover the following:

1   The team will be taken nine-step through the focus on a specific machine, equipment or process TPM improvement plan on their pilot equipment.
2   The team will determine the reliability, performance and maintainability issues. They will put handles on these issues and measure and quantify them using OEE and P-M problem solving analysis.
3   The team will systematically assess and target the major losses as follows:
    - availability:
        elimination of breakdowns (including technique of asking why five times)
        minimum setup and adjust (using SMED and good old method study techniques)
    - performance rate:
        zero idling and minor stoppages
        no running at reduced performance (both goals sought through the continuous improvement habit)
    - quality rate:
        maximum yields (using error-proofing techniques and the CAN-DO or five Ss philosophy of cleanliness and tidiness)
        Minimum startup losses.
4   The team will determine what can be done by whom within a timescale. They will set up a monitoring and evaluation system linked to the on-the-job visibility controls.

5   The team will also decide workplace organization, best practice routines, equipment recording and measurement of overall equipment effectiveness.

When the initial pilot project cycle is completed the team will make a presentation showing what they have implemented and what is still outstanding with respect to:

- standard practice operating and asset care instructions (i.e. best practice routines) with the use of sketches and photographs
- visibility controls with the use of simple graphics, before and after photographs, plus failure records
- potential improvement areas as an assessment of the relevant six losses and the potential for reduction/elimination
- milestones and detailed forward programme
- condition appraisal method and progress including:
  refurbishment programme
  spares and labour requirements
  future planned, preventive maintenance schedules
  daily, weekly, monthly cleaning, inspection and check schedules with clearly identified responsibilities as to who does what
- equivalent OEE measurement method and cost/benefit achievement and future targets linked to timescales.

A typical TPM pilot structure is shown in Figure 7.15.

## 7.8   Roll-out

This is the strategy for extending TPM beyond the pilot project to the whole plant (see Figure 7.14). Figure 7.13 shows a typical timescale for all the activities from awareness to roll-out and beyond. The roll-out strategy is shown in Figure 7.16.

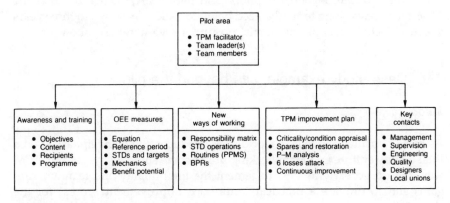

**Figure 7.15** *Pilot structure*

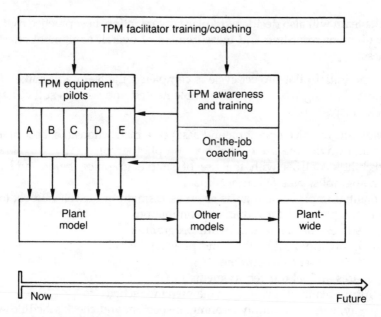

**Figure 7.16**    *Roll-out strategy*

An essential part of the strategy is to assess the time which must be devoted to the TPM process and to prepare and agree a budget so that the necessary funding is made available. The budget must provide for training, team problem solving meetings, and programmes for release of equipment for refurbishing. Refurbishment is often imagined to be expensive, but in reality many equipment restoration activities are no cost/low cost and the first place to look for funding this, plus the required spares, is the existing maintenance budget. All this is part of planning, organization and control, the essential factors of which are discussed in Chapter 8.

There is no better way to illustrate the roll-out phase of TPM than to quote from an actual case study. WCS International started working with this case study company in February 1993. By July 1993 the plant was at the stage of embarking on 'life beyond the pilots' and had clearly recognized that the TPM process was long term, the need being to look at least two to three years ahead. Selected parts of this clients' post-pilot review report follow.

## 7.9    Case study example: Life beyond the pilots

*Introduction*

As part of its intention to be a world class producer, the plant embarked on a total quality culture and teamworking programme some two years ago. TPM was introduced early in 1993 to cement the total quality and teamworking gains and to provide a practical continuous improvement process for the shop floor.

After some four months of TPM experience the plant is able to point to some solid successes through eight initial pilot projects and the start of a plant-wide clear and clean exercise. The basics of a TPM policy and procedure together with a site roll-out strategy have also been defined.

The time is now right, therefore, to take stock and briefly review what has been achieved, but more importantly to look forward and define the TPM vision and the steps for life beyond the pilots. The purpose of this report, therefore, is to:

- review progress to date
- specify the key success factors for the future
- clearly set out the next steps in the TPM process
- highlight those areas requiring management attention and sustained commitment for the future.

### Progress to date (July 1993)

Since the initial TPM awareness sessions and the scoping study were carried out in February and March 1993, a great deal has been achieved. In fact, it is worth recording here the key events and milestones as follows:

*February*
- initial discussions with WCS

*March*
- initial TPM awareness sessions
- scoping study and presentations

*April*
- project mobilization planning
- TPM facilitator training course
- shift team awareness sessions
- key contact sessions
- eight pilot team launches
- TPM steering group review 1
- OEE study started on line 6

*May*
- eight pilot team coaching sessions
- plant clear and clean planning
- TPM steering group reviews 2 and 3

*June*
- TPM policy and procedure draft
- TPM factory notice board in place
- plant clear and clean launched
- pilot team coaching continues
- TPM steering group review 4
- first two TPM team presentations
- TPM study tour to Japan

*July*
- other six TPM team presentations
- TPM facilitators meeting

- roll-out plan developed
- communicate pilots best practice
- continue with plant clear and clean.

Without doubt, the highlight so far has been the eight pilot team presentations after completion of the initial twelve-week TPM improvement plan cycle. These presentations were characterized by:

- significant visual impact improvement on the actual TPM pilot equipments together with their highly effective TPM notice boards
- comprehensive and detailed analysis and review of the pilot equipment, together with the subsequent refurbishment and future asset care and planned maintenance
- an initial assessment of the pilot's overall equipment effectiveness, together with identification of the chronic and sporadic problems and their actual and potential solutions
- definition of the equipment output, fault recording and operator training requirements
- variable but nonetheless conscious efforts by the TPM pilot team's key contacts and squad support members to get involved and contribute to the TPM improvement process
- highly visible and verbal support at the presentations from senior managers and executives of the company
- the generally high-quality presentations given by the eight pilot teams
- an obvious enthusiasm for the TPM process as expressed by the eight pilot teams.

Perhaps this last characteristic is the key indicator for the future success and impact of the TPM process at the plant: namely, that TPM is a grass roots process which the shop floor can *value*.

### Key issues and learning points

Before discussing the next stages and the roll-out of the TPM process, we believe it is *vital* to appreciate, understand and then commit to some fundamental points:

1   The excellent work of the eight TPM pilot teams in terms of the development of best practices must be adopted and adhered to by all shifts. Business team leaders must therefore take this communications responsibility with support from the pilot TPM team leaders, unit managers and TPM facilitators. This is an urgent priority.
2   We cannot manage improvements unless we have clear reference points (this is what it was like), a measurement process (the OEE), and targets (best of the best and world class). We *must* therefore have OEE measures developed and implemented for each of the four lines, reclaim, customer slitting and the plant as a whole. This is also an immediate and urgent priority.
3   The plant clear and clean principles laid down as *eleven* steps (each with a specific purpose) must be pushed along and sustained by managers

and business team leaders to involve *everybody*. This exercise is not an adjunct to the TPM process: it is a fundamental part in order to highlight, reduce and then eliminate causes of accelerated deterioration and dust and dirt, the number one enemy of consistent quality product at the plant.

4   How do we sustain the TPM process when the resources of time and people are already stretched to the apparent limit? The answer is rhetorical, but we have to say to ourselves: 'If we don't do things right in the first place, how on earth do we find time *now* to put things right?' We must convince ourselves of the need to find the time to make it easy to do things right and difficult to do things wrong. The TPM pilot teams are already convinced, so we must encourage them to be the missionaries to spread the message.

5   We need to be thinking about the fact that as TPM develops, the two factory managers will want more control over the day-to-day maintenance of their lines. Conversely, there will be always a need for some central maintenance. It will be necessary to address the following questions:

- What specific tasks are there?
- Which are best decentralized to the business/factory unit?
- Which should stay centralized?
- What resources are needed to support these tasks?
- What transfer (if any) is required and when?

This is a major issue and one which needs careful planning and much thought.

6   Finally, the plant has only just started out on its TPM journey. This journey is a long-distance one and is not a quick sprint. We therefore need a clear TPM *vision* and *process* for getting there. Moreover, the TPM journey will only be completed if the baton of responsibility is clearly handed on from consultant to facilitator, and from facilitator to management and business team leaders. The empowered shift teams will only thrive on TPM provided it has the visible support of the day-to-day management of the plant. The steps to achieve the vision and process are set out in the next section.

### Future action

Our experience of helping other companies to introduce TPM has highlighted several key success factors and *underlined* one specific point:

- The TPM process, whilst gaining significant early benefits, is essentially a long-term process of typically two years plus.

This being the case, we need a clear *vision* and *process* for getting there: 'life beyond the pilots', if you wish.

### The plant TPM vision

Shown in Figure 7.17 is the TPM vision for the plant to achieve a trouble-free plant by striving for zero targets through teamwork, communication and

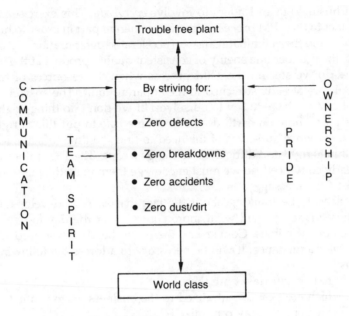

**Figure 7.17**   *Factors in world class performance*

pride of ownership to achieve world class levels of overall equipment effectiveness.

## The TPM win commandments

The ten win commandments shown in Figure 3.12 must be adhered to in order to support and achieve the TPM vision. As such, it is vital that *everyone* at the plant reads, understands and is committed to each of the ten statements. Equally, they are just hollow statements unless we actually put them into practice.

### Four phases of the TPM improvement process

As we have stressed throughout the initial eight-pilot introductory phase, the TPM process is not completed within the twelve-week period. Rather, it represents the completion of phase 1 in Figure 7.18. In other words, as rollout progresses, each of the initial pilots and their subsequent pilots are revisited over four distinct phases:

1   introduction                                    3–6 months
2   refine best practices and standardize           6–12 months
3   build capability                                12–18 months
4   continuous improvement                          18–24 months.

| | Condition cycle | Measurement cycle | Improvement cycle |
|---|---|---|---|
| *Phase 1*<br>Introduction<br><br>3 to 6 months | • Refurbishment<br>• Planned maintenance<br>• Identify accelerated deterioration | • Equipment history<br>• Set best of best targets | • Known problems<br>• No-cost and low-cost solutions<br>• BPR proposals<br>• Visual help |
| *Phase 2*<br>Refine best practice and standardize<br><br>6 to 12 months | • Identify design weaknesses<br>• Stop accelerated deterioration<br>• Forward action plan<br>• Maintainability issues | • Early warning systems<br>• Quantify link to commercial drivers<br>• Achieve consistent best of best | • Visual management<br>• Improve methods<br>• Support solutions<br>• Quantify technical problems |
| *Phase 3*<br>Build capability<br><br>12 to 18 months | • Eliminate accelerated deterioration and sources of dust and dirt<br>• Establish optimum conditions<br>• Train and develop skills<br>• Forward training plan | • Set zero targets<br>• Forward business plan<br>• Performance improvement<br>• Quality of maintenance | • Define hidden problems<br>• Technical solutions<br>• Integration structure and systems |
| *Phase 4*<br>Continuous improvement habit<br><br>18 to 24 months | • High-level poke yoke<br>• Designed in maintenance prevention<br>• Equipment strategy plan implemented | • Strive for zero targets<br>• Commercial, bottom line results | • New Technology<br>• Improved equipment design performance<br>• 'Better than now'<br>• Flawless operation |

**Figure 7.18** *Four phases of TPM implementation*

via the condition, measurement and improvement cycles. Adherence to these four phases is fundamental to the achievement of TPM as a 'way of life' at the plant.

The site roll-out proposal is shown in Figure 7.19.

## 7.10 Training and awareness

Figure 7.20 provides a graphic indication of the learning process. The low retention rate when simply reading or listening to what others have to say is illustrative of 'in one ear and out the other'. The rate improves when there is a visual aid to what is being said, and progresses further when there are opportunities for discussion. Real progress comes with the opportunity to practise and experience what we are taught – and we are almost home and dry when we reach the level where we can teach others!

Effective training is absolutely essential if the TPM process is to be carried through to a successful outcome. One approach is single-point lessons, which are confined to a single activity or learning point and consist of explanation, demonstration, practice and confirmation. Properly applied these lessons can lead to a high retention of knowledge and skill, developing to full competence and in due course the ability to train others.

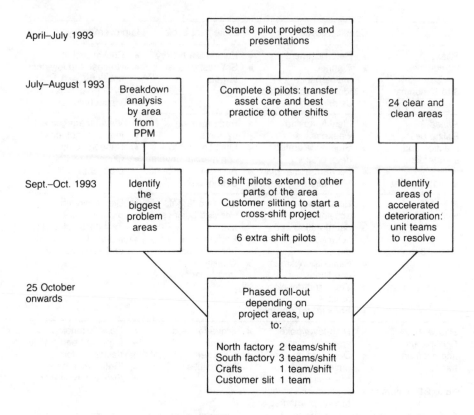

**Figure 7.19**   *Site roll-out plan proposal*

---

      5% of what we read
      10% of what we hear
      15% of what we see
      20% of what se see and hear
      40% of what we discuss with others
      80% of what we practise and experience
      90% of what we teach others

---

**Figure 7.20**   *Learning and understanding*

Other approaches with a broader base leading to more comprehensive understanding of more complex activities are CD-i media, videos, slide shows, written material for study, and illustrations.

Figure 7.21 summarizes all the principal learning points associated with TPM, and uses the shaded quadrant approach to indicate progress made in the training/learning process. Figure 7.22 focuses on the key learning points, all of which have been discussed and described in this and earlier chapters. Figure 7.23 portrays a training cascade by means of which the TPM messages are spread progressively through the whole organization.

Name _____     Company _____

Date _____

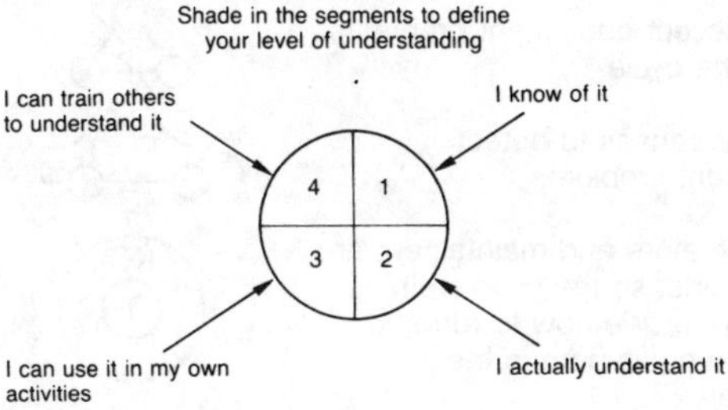

Shade in the segments to define
your level of understanding

I can train others
to understand it

I know of it

4 | 1

3 | 2

I can use it in my own
activities

I actually understand it

1   The TPM iceberg

2   Don't live with equipment problems

3   The asset care reality (apple a day etc.)

4   Maintainer/operator roles and communication

5   Key contact roles and communication and job titles

6   Using our senses to detect problems: car analogy

7   The benefits of TPM

8   TPM is not rocket science, but common sense

9   The six losses

10  CAN-DO and visual management

11  Objectives of pilots

12  OEE

13  Criticality assessment

14  Condition appraisal/refurbishment

15  Asset care regime

16  History recording

17  Problem solving and the six losses

18  Best practice routines

19  Reducing accelerated deterioration

20  Improving the quality of maintenance

21  Passing on lessons learned from exercises

22  Teamwork

23  Resistance to change

24  Creating the right environment

**Figure 7.21**  *TPM learning points*

1 We need to be aware of the hidden cost caused by the quick-fix approach to equipment problems

2 Don't accept equipment problems: break the cycle

3 Use our senses to detect equipment problems

4 Our operators and maintainers are the experts, so they can (with training) decide how to run and maintain equipment in the best way

5 We must improve teamwork between them

6 Together we can *measure* the benefit, set and maintain optimum *conditions*, and continuously *improve* our equipment

Measurement cycle

Condition cycle

Improvement cycle

**Figure 7.22**   *Key TPM learning points*

| Typical numbers | Group | Content |
|---|---|---|
| | *Coach* | |
| 1 to 15 | 1 TPM facilitator | • 4-day facilitator course |
| | *Teams* | |
| 15 | 3 TPM team leaders and teams | • 1-day TPM training (all 3 teams)<br>• 9-days on-the-job coaching (3 each) |
| | Repeated as new pilots are set up by facilitator | • 1-day recap and presentation (1 for each team) |
| | *Key contacts* | |
| 40 | Other team leaders, foreman, managers, union reps, QA, personnel, MPC, Med., financial | • 1-day TPM awareness course (3 repeat sessions) |
| | *Plant/site* | |
| 170 | All other production and maintenance team members | • 1/2-day TPM awareness (6 repeat sessions) |

**Figure 7.23**   *Training cascade*

Figure 4.8 charts the progress through training from innocence to excellence. Figure 4.9 is particularly interesting because it portrays what happens to us when we learn routines so thoroughly that we do not need to think when we are performing. Human life is full of situations where unconscious competence is taking place. This is what training aims to achieve.

Some training and awareness objectives in TPM are:

- not just 'how' but also 'why'
- resolution not just fixing
- greater skills development and flexibility
- increased diagnostic, recovery abilities
- responsibility for plant, process, equipment condition.

In short, team working means harmony and partnership: maintenance must be productive, whoever does it.

A typical training programme spread over a four-month period is shown in Figure 7.24. WCS International's comprehensive training and development plan is reproduced later in this chapter.

## 7.11 Exit or completion criteria

This is equivalent to the end-of-term examination in which the achievements and lessons learned in a given phase are checked. Only if these exit criteria are satisfied can the group proceed towards the next objective. The concept is illustrated in detail in Figure 7.25.

A typical page from an exit criteria document is shown in Figure 7.26. This page is taken from a comprehensive approach and includes, in addition to OEE, criticality and condition appraisal, further check sheets covering refurbishment, asset care, equipment history details, assessment and scope of equipment losses, problem solving and best practices.

| | 20/6 | 29/6 | 30/6 | 6/7 | 7/7 | 13/7 | 14/7 | 20/7 | 21/7 | 17/8 | 18/8 | 24/8 | 25/8 | 7/9 | 8/9 | 14/9 | 15/9 | 21/9 | 22/9 | 28/9 | 29/9 | 5/10 | 12/10 | 13/10 |
|---|---|---|---|---|---|---|---|---|---|---|---|---|---|---|---|---|---|---|---|---|---|---|---|---|
| TEAM 1 | ■ | | ■ | ■ | | | | ■ | □ | | | | □ | | | | | | ■ | □ | □ | | ■ | |
| TEAM 2 | ■ | | | | ■ | □ | | | | ■ | ■ | □ | | □ | ■ | | □ | | | | ■ | | | |
| TEAM 3 | ■ | ■ | | | | | □ | ■ | | | | ■ | | | | □ | ■ | | | □ | | | ■ | |

| FOR ALL OTHER TEAMS MEMBERS | 18/7 | / | 22/8 | 12/9 | 28/9 10/8 | 3/9 | 29/8 | | | | | 4 HR GENERAL AWARENESS TPM SESSIONS  1 SESSION PER PERSON | | | | | | | | | | | | |

| | 26/6 | 21/8 | 28/8 | | SQUAD TRAINING ONE DAY SESSIONS TPM  T.U. M.P.C. Q.A. FINANCE ENGINEERING  TEAM LEADERS FOREMEN SHIFT MANAGERS  SENIOR FOREMEN MECHANICAL ENGINEERING | ONE-DAY ON THE JOB SESSIONS ■ WITH THE CONSULTANT'S ASSISTANCE |
|---|---|---|---|---|---|---|

PROBLEM SOLVING COURSES FOR THE TEAMS    TEAM ONE    30 Aug. 1, 2 Sept
    TEAM TWO    26, 27, 28 Aug
    TEAM THREE    16, 17, 18 Sept

**Figure 7.24** *Typical TPM training programme*

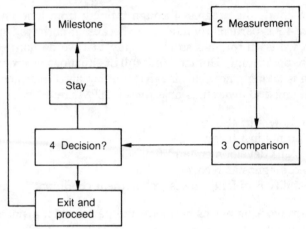

**Figure 7.25** *Milestones, measurement and exit criteria*

Milestone 1: TPM Pilot
Exit criteria, phase 1 TPM improvement plan

| | Measurement | | |
|---|---|---|---|
| | Yes | No | Date due |

*1 Overall equipment effectiveness*

Have all the equipment losses been identified?

Have all the losses been quantified under the headings of availability, performance rate and quality rate?

Have average, best of best and ultimate standards of OEE been established?

Does the OEE reflect a true measure of production versus capability?

Has the benefit of a 1% increase in the OEE been evaluated?

*2 Criticality assessment*

Have all the components which may impact on the assessment variables been identified?

Do all members of the team agree on the relative impact of each component of these variables?

Do other operators, maintainers and supervisors agree on this ranking?

*3 Condition appraisal*

Has a general statement of reliability and maintainability been made?

Has the condition of all the components listed in the criticality assessment been evaluated?

Has the condition of other components been evaluated?

Have photographs been taken to support these statements?

Have the causes of accelerated deterioration been identified?

**Figure 7.26** *Example page from exit criteria document*

Figure 7.27 shows a TPM review summary format which assesses TPM on the broader but essential front of the business driver and business strategy deployment activities.

## 7.12 WCS International TPM training and development programme

WCS International's unique TPM training and development packages are based on practical training and implementation experience in:

- over 60 plants
- in 40 different manufacturing, process and utility industries
- in 12 different European countries.

### The WCS approach to TPM

In helping our customers to introduce TPM principles, philosophy and practicalities into their company, we have developed a unique and structured step-by-step approach based on the TPM implementation route shown in Figure 7.1.

1  An initial half-day to one-day general TPM awareness session for senior management, supervision, employee and employee representatives.
2  A short, sharp scoping study to determine how TPM can be tailored to suit the specific plant situation. This is typically based on initial TPM pilots linked to a site roll-out programme.
3  A mobilization phase to agree to commitment, delivery, site awareness and detailed training for the TPM pilots and subsequent site-wide implementation.
4  Plant-specific TPM awareness module for all employees directly or indirectly involved in TPM.
5  A comprehensive and practical four-day TPM facilitator and supervising training course.
6  Launching and supporting the TPM pilots including comprehensive TPM awareness sessions and training of management and support staff, team leaders and team members including extensive and practical on-the-job coaching.
7  Parallel corporate and plant-level project management support to ensure a coherent and consistent TPM policy, methods, publicity, standards and measurement process with clear completion criteria.
8  Site-wide roll-out planning and implementation based on a continuous improvement culture.

In our experience it is *vital* to tailor your TPM implementation plan, not only to suit our differing Western cultures and industry types, but also to recognize the sensitivity of local plant-specific issues and conditions.

Assessment by:      Location:      Date:

| Area | Vision | Assessment | Action |
|---|---|---|---|
| Business management/ strategy | Management understand the value of hidden costs and the six losses. They know what TPM adds to their strategy and how to use the benefits of increased capacity, reliability and flexibility to support the achievement of world class performance and maintenance of zero targets. Equipment management is a coordinated function from initial design concepts through the complete equipment life cycle. | | |
| Infrastructure and technology | Technology issues are understood and under control. There are clearly defined equipment management roles and responsibilities. All departments play a positive role in improving equipment effectiveness. TPM is integrated into the normal fabric of management. | | |
| Systems in place and working | Routine maintenance is carried out autonomously. Lifetime maintenance systems are in place for all equipment. Lifetime cost information is available and accurate. Visual management is in use on the machine, for performance trends, training aids and to support flawless operation. Effective problem solving and equipment early warning feedback to design are in place. Standard procedures used, widely recognizing the principles of CAN-DO including the use of shadow boards and lineside stores. Appropriate condition monitoring in place. | | |
| Objective feedback/ communication | Clear priorities are set by management and translated from the management vision into departmental goals in support of maintaining zero targets and world class OEE levels. Rewards support the application of TPM concepts such as routine maintenance and solving problems for good. | | |

Skill development — Personnel are encouraged to enhance skills and improve ability to support the use of TPM concepts.
A systematic approach is used to departmental training needs definition and skill development.

Cultural indicators — Cross-departmental small-group problem solving activities are autonomous.
Interest in finding out more is stimulated by simple models and sketches.
Maintainers treat breakdowns as an opportunity to train operators.
There is a clean, bright working environment.
TPM boards and visual methods are used effectively to communicate goals and performance results.

Motivation — There is a positive response to new ideas and high equipment ownership at all levels throughout the organization.

Action strategy:

**Figure 7.27** Review summary

## TPM awareness session

### Objectives

The objectives of the session are threefold:

- to give an awareness of TPM and how in practice it can fit with the company's current progress and future vision and values
- to develop the basic building blocks usually within the existing TQM culture for a future plant TPM programme
- to agree a way forward.

### Structure and participation

We split the presentation time and open forum time on a roughly 70/30 basis and stimulate the open forum session with, perhaps, three key questions or learning points to be addressed or worked through in syndicates.
   Typical questions might be:

- What benefits do we believe we would get from TPM for the business?
- Which area/process would make a good pilot and why?
- What would be our key success factors for TPM?

This session can be designed to run for two hours, a half-day or a whole day. Attendance numbers can vary from 6 to 30 plus, with about 20 being optimum.

## Scoping study

### Objectives

The objectives of the scoping study are to tailor your TPM implementation plan, not only to suit our differing Western cultures and industry types, but also to recognize the sensitivity of your local, plant-specific issues and conditions.

### Content and format

Typically, over a one-week to two-week period, we will work with you to position the variables of:

- people's perceptions and feelings
- organization structure
- commitment to change
- pace of change
- status of change
- equipment condition, type
- processes and types
- measurement
- potential benefit
- priority and integration.

Together, we can then tailor your TPM implementation and training plan accordingly.

## Outputs

The outputs from this initial short, sharp scoping study will be an agreed presentation document and report which will cover:

- a quantified and agreed assessment of equipment losses and potential benefit
- people's perceptions and the implicit change process
- training and implementation for TPM:
    key success factors
    potential pilots
    team size and membership
    squad membership and roles
    logistics and resources
    initial awareness and training
    content and timing plan
    TPM facilitator support
    TPM steering group
- suggested roll-out plan for site.

### Mobilization phase

## Objectives

Following the scoping study, the objectives of this phase are:

1   To set up the actual teams and the support of the initial pilots including actual names of the TPM teams, key contacts, squad support members, drawn from production and maintenance for the teams, and from quality, production control, finance, purchasing, design and engineering for the squad, as well as business and unit managers and supervisors as key contacts.
2   To define and agree the detailed time phasing of the training programme: dates, times and attendees.
3   To set up the TPM steering group and meeting dates to review progress.
4   To set up the TPM facilitator support resource, including date of facilitator training.
5   To ensure clear expectations, understanding and commitment to the planned process from all concerned.

## Timing

This phase should only take two to three days to complete, spread over a week.

*Specific TPM awareness sessions*

## Objectives

On completion of the scoping study and mobilization, it is now possible to produce a plant-specific TPM awareness training module to be presented to *all* employees who are going to be directly or indirectly associated with the TPM process.

The module will last about two hours. The objectives are to answer three questions for employees:

- What is TPM?
- Who is involved?
- How will it affect me, and when?

*Facilitator training*

## Objectives

The objectives of the four-day in-company course (Figure 7.28) are:

- to provide a thorough understanding of TPM and how to put it into practice in your company
- to provide a framework and understanding for TPM facilitators to work in and to influence the behaviour of multi-disciplinary, multi-interest teams.

## Format

At least 70% of the course is focused on carrying out the TPM process on your own live TPM pilot equipment on-site so that delegates can experience the on-the-job reality of putting TPM into practice.

## Who should attend?

*Proposed TPM facilitators*
*TPM champions*   You may be the managing director, general manager or similar person with executive responsibility.
*TPM driver*   You will be a middle manager, supervisor or team leader at the sharp end of the business, responsible for producing quality output from consistently highly reliable equipment.

## What will you learn?

- By the end of the course you will have a thorough understanding of TPM.
- You will be equipped to set up the TPM planning, process and delivery for your successful in-house TPM programme.
- You will have a wealth of relevant training material to help your company on its TPM journey.

| Day 1 | Day 2 | Day 3 | Day 4 |
|---|---|---|---|
| • Introduction to TPM <br> • Ex. 1: Whose job? <br> • TPM principles <br> • Slide presentation | • Recap brief <br> • Visit to pilot and plan the plan <br> • Ex. 5: Criticality assessment <br> • Condition cycle (on-the-job) <br>     Criticality assessment of pilot <br>     Equipment description | • OEE for pilot <br> • Consolidate condition and <br>     measurement cycles <br> • Supporting techniques <br>     P-M analysis <br>     six losses <br>     assessment | • Dry run presentations <br> • Syndicate presentations <br> • Review and key learning points |
| • TPM techniques <br> • Ex. 2: OEE <br> • Ex. 3: Project management <br> • Teamworking and facilitating <br> • Ex. 4: active listening <br> • Briefing for syndicates <br> • On-the-job case study | • Condition appraisal <br>     Refurbishment plan <br>     Asset care <br> • Measurement cycle <br>     (on-the-job) <br>     History/records <br>     OEE measures | • Improvement cycle <br>     (on-the-job) <br>     Assess losses <br>     Problem solving <br>     Best practice routines <br> • Prepare presentation <br> • Ex. 6: Recap | • Ex. 7: Whose job? <br> • Pilots and roll-out <br> • Supporting the pilots <br> • Course assessment |

**Figure 7.28** *Facilitators' course programme*

*Team and squad support: training for pilots and roll-out*

## Objectives

The purposes of using pilot projects to kick off the TPM process are many and varied, and will include the need to:

- provide a lever for change
- demonstrate the power of TPM
- prove the TPM improvement plan
- show effective teamworking
- objectively test the water
- address hindrances
- build a TPM policy and procedures bible
- plan for roll-out and infrastructure
- provide a solid, real example on the basis that seeing is believing
- provide an enabling process for management and their employee representatives (unions or works councils).

The objective is therefore to focus on a particular TPM pilot or pilots, not only as a means for gaining substantial and measurable improvements in overall equipment effectiveness, but also as a showpiece and template for the on-the-job work that is really at the heart of TPM. This is achieved with some classroom training, but the emphasis is on the on-the-job coaching aimed at assessing and then eliminating the six big losses against a reference period calculation of the OEE for that particular pilot equipment, process, area or machine.

## Overall training programme

Select the TPM pilot(s) to be the focus of a centre of excellence.

Carry out a one-day in-depth session of TPM training, as opposed to TPM awareness, which concentrates on the WCS TPM improvement plan. This session must be attended by *all* TPM members and their key contacts and squad support members. The day is completed with a crystal-clear set of objectives and messages, for example:

- Over the next $x$ weeks (typically 12) we are going to start to implement these principles and techniques of TPM, and in particular the condition, measurement and improvement cycles of TPM back on the pilot area.
- This means that in 12 weeks' time we will come back here and, as a team, present our achievements, plans and intentions.

During the ensuing period, back on the TPM pilot, the team will be coached in the hands-on application of TPM. This will be aimed at a coherent focus for specifying, developing and then implementing a TPM programme to give measurable financial benefit in terms of OEE performance. Also a highly visible TPM pilot will give encouragement to others on site who are starting out on the TPM journey.

A typical TPM pilot training programme is illustrated in Figure 7.29.

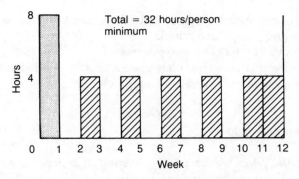

Based on TPM equipment teams of about 5 people

| Week 1 | 1 day TPM awareness and training |
| Weeks 2, 3 | OEE evaluation, criticality assessment, condition appraisal |
| Weeks 4, 5 | Refurbishment, asset care, recording |
| Weeks 6, 7 | Assessment of the six losses |
| Weeks 8, 9 | Problem resolution, best practice routines |
| Weeks 10, 11 | Prepare for presentation |
| Week 12 | Presentation |

**Figure 7.29**   *Typical pilot commitment for each member of a TPM team*

## Plant-level TPM project management support

Based on our experience of working in over 60 sites and plants where TPM is being pursued, we can provide support in the following areas:

1   Development of corporate and plant-level TPM policy and procedures statement and work book covering the following topics:
   - introduction and aims
   - company's beliefs and values, TQ and TPM
   - corporate and plant-level TPM process
   - TPM key success factors for the plant
   - site TPM policy and procedures statement
   - key milestones, measurement and completion criteria
   - TPM for design and engineering
   - TPM training modules
   - TPM publicity.

2   Provision of TPM audits and reviews and support to the in-house TPM steering group whose objectives are:
   - to ensure a consistent and measurable introduction of TPM

- to regularly review progress and achievements against agreed milestones and completion criteria
- to remove any blockages to progress
- to ensure that progress, plan intentions and publicity material are regularly communicated to *all* employees.

### Roll-out programme support

Beyond the initial pilots we can provide assistance in defining and supporting the TPM roll-out programme for the plant.

This roll-out programme definition is unique to each plant in terms of its scope, content, timing and resourcing. It can usually cover an implementation period of two to four years before TPM becomes a key part of the plant culture and continuous improvement. The TPM roll-out process can be illustrated as in Figure 7.30.

We can deliver training for

- operators
- maintainers
- supporting function
- management
- facilitators/supervisors

to five levels of competence, namely where the employee:

1  is aware of and recognizes the value of TPM
2  understands how TPM is applied
3  has used the concept and techniques of TPM
4  has achieved defined levels of success using TPM
5  promotes the use of TPM as a continuous improvement habit.

The aim of the programme is to get all personnel to level 5.

The four phases of the improvement process have been shown in Figure 7.18. Within each of the four phases, key milestones are achieved, and a typical illustration is given in Figure 7.31.

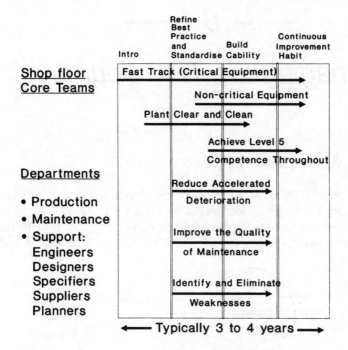

**Figure 7.30** *Plant roll-out of TPM improvement process*

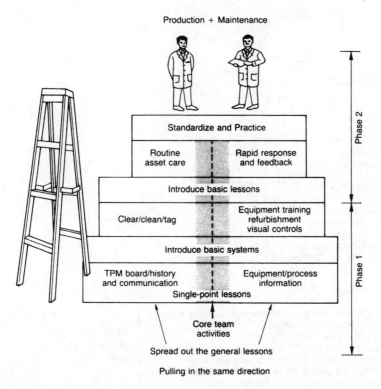

**Figure 7.31** *Refining best practice: phase 2 milestones following phase 1 introduction*

# — 8 —

# Managing the TPM Journey

## 8.1 Future vision, planning and control

The introduction of TPM to an enterprise starts with a vision of the future, and this is illustrated in very clear terms by Figure 8.1. All the means of achieving TPM which have been discussed in earlier chapters lead to the continuous improvement habit, which embodies the spirit of Kaizen and

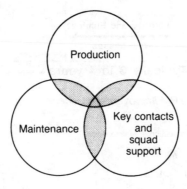

*Soft benefits*

- Recognition
- Teamworking
- Harmony
- Partnership
- Motivation
- Winning
- Helping one another
- Ownership
- Pride
- Self-esteem
- Problem solving
- Problem resolution
- Wanting to change
  the way we do things

} Continuous improvement habit

*Hard benefits*

- Improved OEE gives choice not currently enjoyed
- Shortened lead times, less WIP/RM
- Reliability of customer promise
- Quicker response to new customers and needs
- Overall perspective to elimination of waste
- Fast track priority for initial bottleneck pilots
- Ensure past constraints remain 'debottled'

} World class, lean producer, working just in time

**Figure 8.1** *TPM as an aid to future vision*

Nakajima and which can be brought to reality by following the WCS approach to Western TPM. The key point is that when people *want* to change the way they do things, then they will, and they will sustain it.

Some of the major changes which will result from the introduction of TPM and the benefits which those changes will bring are as follows:

| | |
|---|---|
| Machines run close to name-plate capacity | Reduce need for excess capacity |
| Ideas to improve often proposed by operators | Ownership/success |
| Breakdown rate reduced | Used to learn and teach the team |
| Machines adapted to our need by our people | Our machines will be better |
| Operators solve problems themselves | Fewer delays and stoppages |
| Cleanliness and pride in continuous improvement | Good working environment |
| More output from existing plant | More profits |

Planning, organization and control are essential prerequisites:

*Planning* entails allocation of resources on a realistic and achievable basis with regular review and progressive development on the long-term basis necessary for success.

*Organization* requires defined resources with clear allocation of roles and responsibilities; this must be accompanied by effective and clearly understood methods of working.

*Control* The two aspects of control are *coordination*, which is concerned with what happens next and is most effective with simple visible systems and procedures; and *feedback*, which is

- concerned with goals for time, cost and quality
- used to identify the reasons for failure and to prevent recurrence
- the source of objective evidence of the need for increased resources, modification of goals or the introduction of specialists.


## 8.2 Role of management

Critical success factors for management may be summarized as follows:

*Planning*
- Ensure sufficient standard process and measurement.
- Ensure realistic understanding of expectations.
- Plan for consistent criteria of measurement and recognition.

*Process*
- Maximum involvement from entire organization.
- An effective TPM communication.
- Clear exit criteria before roll-out.

*Support*
- Detailed training and technical information.
- Sufficient and quality manpower to work on TPM.
- A budget of money *and* time.

## 8.3  Cost/benefit profile of TPM

Introduction of TPM to a company which has relied heavily on breakdown or reactive maintenance inevitably entails initial additional time and training costs. As the organization moves away from reactive maintenance towards the proactive or planned, preventive approach, and as operators play an increasing role, the benefits will show up and yield significant permanent savings in costs.

This effect is well illustrated by Figure 8.2. A typical break-even point is likely to occur after about one year, with the real permanent reduction in costs likely to be realized by the third or fourth year. To questions about the initial increased cost of TPM, a typical Japanese reaction would be that they had given it scant consideration because their eyes were firmly fixed on the permanent long-term advantages, not just in maintenance costs but in the lost opportunity costs of a low overall equipment effectiveness.

Figure 8.2 combines the bow-wave effect with the increase in OEE, and shows clearly how initial costs are rapidly converted to permanent benefits.

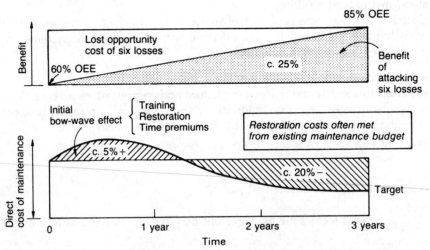

**Figure 8.2**  *Typical TPM cost/benefit profile*

Experience shows that a 5% increase in the OEE is often equivalent to at least 25% of the annual maintenance cost.

## 8.4  Steps to achieve the TPM vision

The factors which drive the TPM process forward are as follows:

1   measurable financial benefit
2   top-down target-driven management
3   bottom-up shift team activity:
    - leading change
    - linking change
      delegation
      devolution
    - Coherence
      targets
      responsibilities
      training
      wanting to change
    - Communication
      understand problems
      better, positive response
    - Environment
      pride in workplace
4   teams of capable, creative, self-sufficient people.

The complete four-phase process from pilot scheme to stabilization of TPM is illustrated in Figure 8.3. This shows how the driving force moves progressively from the coach or facilitator through to the maintainer and operator. It also shows clearly the role and involvement of production, maintenance, supervision and management in the various phases of development.

## 8.5  Management structure and the roles of supervisors

A plant-level TPM project structure taken from an actual case history is shown in Figure 8.4. This shows clearly the involvement of top management and the relationship of all the aspects of TPM which have been addressed in earlier chapters.

The roles and responsibilities of the management and supervisors as key contacts to the TPM teams were mentioned in section 7.5 and are repeated here to underline their importance:

- input and release of people for TPM training
- release of equipment for restoration and subsequent asset care
- technical and historical information
- plant and line OEEs
- commercial and market benefits definition

| Introduction (pilots) | Refine best practice and standardize (roll-out) | Build capability (promotion and practice) | Continuous improvement (stabilization) |
| --- | --- | --- | --- |
| Facilitator driven | Facilitator/supervisor driven | Supervisor driven | Small-group maintainer/operator driven (i.e. self-sustaining) |
| *Production focus* Selected operators in pilot area work on improvement plan | All departments begin improvement activities | All employees use TPM concepts Establish standardization | Autonomous small groups implement activities Early problem detection/solving |
| *Maintenance focus* Selected maintainers work with pilot team | Maintenance organizes to support plant-wide projects and planned maintenance | Training in maintenance skills for operators and maintainers | Reducing equipment lifetime costs |
| *Management focus* Policies and structure to support long-term commitment | Stimulating interest, managing resources | Encouraging teamwork, training and skill development | Striving for world class performance |

**Figure 8.3**  *Steps to achieve the TPM vision*

- visual management of information
- TPM publicity and awareness communication
- inter-shift communications
- TPM activity logistics and facilitator support
- standardization of best practice
- spares forecast and consumption rates
- hygiene and/or safety training and policy
- input to problem solving and solutions support.

A dramatic message for managers and supervisors is embodied in Figure 8.5.

## 8.6  Barriers to introduction of TPM

Inevitably when major changes in an enterprise are being introduced there will be suspicion and opposition: this has been discussed in Chapter 7. Some of the common reasons for suspicion are as follows:

- Management shows impatience for quick fixes rather than stickability and commitment.
- I operate, you fix: I add value, you cost money.
- Operators are taking our jobs away (say maintainers).
- TPM is a people reduction programme (threat of job losses).
- TPM is just another cost reduction driven.

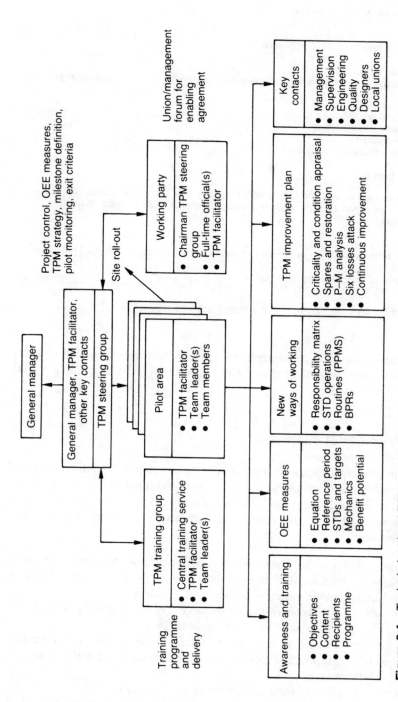

**Figure 8.4** *Typical plant-level TPM project structure*

*Requirement*
To be a world class manufacturer

*Through*
Just in time, lean production, one-piece flow

*Prerequisite for success*
We will be planning for failure regarding our daily and weekly output schedules *unless* the six losses of

- setup and adjust
- idling and minor stoppages
- reduced speed
- defects and rework
- startup
- breakdown

are systematically tackled and eliminated through the TPM improvement plan. This will be reflected in an overall equipment effectiveness performance of 85%+ rather than the c.70% OEE of today.

*The choice?*
- Either 100% sustained commitment from management and supervision for TPM

*or*
- Continue to plan for failure.

**Figure 8.5** *Message from TPM for managers and supervisors*

In the early stages, if communication is poor then resistance will be inevitable. An open and clearly thought out agenda for TPM is absolutely vital.

## 8.7 Visibility of information

The importance of visible sources of information to reinforce discussion and verbal instructions cannot be too strongly stated. One of the major lessons learned in the early stages of TPM introduction in Japanese enterprises was the use of sight as part of communications to complement hearing. Some examples are:

*TPM equipment boards* Handwritten documentation of status, progress, achievements prominently displayed in the work area, with objectives included. A schematic example is shown in Figure 8.6.
*Black museums* Examples of problems solved, or waiting to be solved.
*Training records* Displayed publicly: updated by trainees.
*Notice boards* Located at factory entrances. Professionally implemented, with excellent graphics. Processes and achievements clearly analysed. Include safety records.
*Operation information* Hung just above operators (A4 format).
*Maintenance points* Marked by red arrows and frequency symbols. Colour coding for clarity of assembly/repair sequence. Optimum maintenance routes clearly marked. Machine defects tagged.

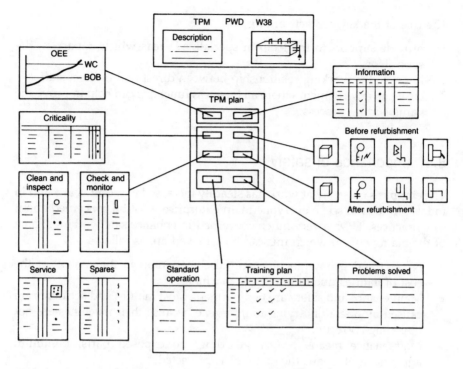

**Figure 8.6** *Example of the visual management of information*

## 8.8 Support for teams and key contacts

The role of teams and key contacts has been discussed at length in Chapter 7. Management support for these groups must be visible and total. Support of teams involves: active listening; supporting and, if differing, taking responsibility; stating issues which are your concern; and being specific. Some negative team practices which need to be gently corrected are: over-talking; not joining in; flying off at a tangent; and hijacking the discussion.

Some guidelines for helping teams to succeed are:

- agreed priorities and strategy
- effective planning, control and delivery systems
- clear organization of labour, equipment and materials
- insistence on measurable results and individual commitment to them
- encouragement to identify and meet task and process skills needs
- active reinforcement of teamworking
- promotion of a positive outlook to problem solving and new ideas
- mutual trust
- mutual support
- good communication
- shared objectives
- managed conflict
- effective use of skills.

The role of the key contacts are to:

1    provide support to the team in specialized areas which impact on the shop floor
2    improve the working relationship between direct and indirect staff
3    create an environment where all business functions can help improve the value adding process.

## 8.9    Importance of safety

The emphasis on safety at work has steadily increased in recent years, and in today's industrial scene everyone in an enterprise must be concerned about safe practices. TPM is very much concerned to enhance safe working. Some of the main ways in which this can be achieved are as follows:

- Neglect and penny-pinching are false economies in the context of the cost of injuries due to unreliable machines.
- Maintenance and safety are tied partners. Most injuries and accidents are caused by operators trying to intervene because their machines are not operating correctly.
- Maintenance means proper guarding, no exposed parts, minimum adjustment: it means the operator is protected.
- The Health and Safety Executive says most hearing damage is caused by badly maintained machines.
- When cleaning or driving our car we can identify at least 27 condition checks, of which 17 have significant road safety implications. Bring this good practice into work with you (see Figure 3.1).
- The notion of the competent and trained person, linked to assets that are fit for purpose and safe, plus statutory obligations, must be central to your TPM strategy, policy and practice.

## 8.10    Summary

Figure 8.7 shows graphically how and why TPM works. Some of the intangible benefits are:

1    TPM is common sense and is therefore valued by employees and employers alike.
2    Vehicle for implementing the company's goals and vision.
3    Change the employees' mind, creating ownership:
- belief in his/her equipment
- protected and maintained by him/her
- through self-help (autonomous maintenance).
4    Give the employees confidence in themselves: create a feeling of 'If there's a will, there's a way.'
5    Clean environment, and environmentally clean.
6    Good corporate image, intangible.

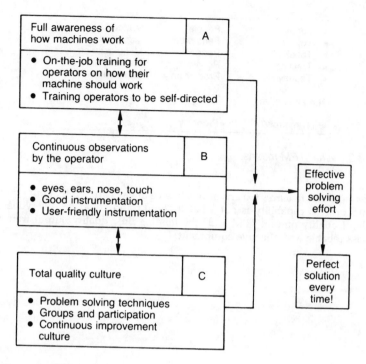

**Figure 8.7**  *Why TPM works*

---

*Question*
If you haven't got the time to do things right the first time . . .
how are you going to find the time to put them right?

*Answer*
TPM gives you the time to do things right the first time, every time!

---

**Figure 8.8**  *TPM: the answer to a problem*

Figure 8.8 poses a question and provides the answer which epitomizes the TPM approach. Figure 8.9 illustrates what TPM meant to a team based on their experiences of running a 16-week TPM pilot exercise at an automotive manufacturer in the north of England.

There is no better way of rounding off this chapter than by quoting the general manager of the plant after attending a team presentation of a TPM improvement plan pilot:

> We started our TPM programme – or TPM journey as I prefer to think of it – about three months ago, so it's early days yet. However, the things that struck me most about the TPM team's presentation today were their obvious enthusiasm for what is proving to be a grass roots process with real business benefits. The other factor which is quite clear to me is that TPM can only be sustained *provided* our supervisors and managers support the TPM process wholeheartedly. Our workforce obviously values the process: it is up to us to give them the

| | | |
|---|---|---|
| • Today | People | Matter |
| • Totally | Pampered | Machines |
| • Totally | Perfect | Manufacturing |
| • Training | People | Meaningfully |
| • Teamwork: | Production + | Maintenance |

The alternative:

| | | |
|---|---|---|
| • Tomorrow? | Probably . . . | Maybe . . . |

**Figure 8.9**   *What TPM means*

*time* and full *resources* to carry it out. We've always known that our equipment and process capability is not what it should and could be. Everyone thinks about quality output. TPM adds the missing link: quality output from world class reliable and effective equipment.

# 9

# TPM for Equipment Designers and Suppliers

Behind the plant and equipment used in the production process there are five functional groups, namely:

- designers
- engineers
- specifiers
- planners
- suppliers.

These five, together with operators and maintainers, make up the seven essential partners in the TPM process. This chapter describes in outline how the activities of the seven must be coordinated and focused on the TPM objectives. The partnership requires a sustained drive towards improving equipment performance through the elimination of the reasons for poor maintainability, operability and reliability.

Designers and engineers need to improve their skills by:

- regular visits to the shop floor and learning from what operators and maintainers have to say
- studying what has been achieved in equipment improvement as a result of self-directed and quality maintenance activities
- gaining hands-on experience with equipment including operation, cleaning, lubrication and inspection
- supporting P-M analysis as part of the key contact/team activities
- conducting maintenance prevention analyses.

Figures 9.1 and 9.2 show how the five goals of TPM can be achieved through design feedback, early warning systems and objective testing of new ideas.

Figure 9.3 portrays the plant roll-out of the TPM improvement process beyond the initial pilots. All seven partners are involved in achieving the continuous improvement habit, typically over three to four years.

Figure 9.4 has appeared as Figure 4.6, and is reproduced again without apology!

## 9.1 Error-proofing

In Chapter 4 various visual techniques were listed. The first steps towards error-proofing would include:

- guide pins of different sizes
- error detection and alarms
- limit switches and stops

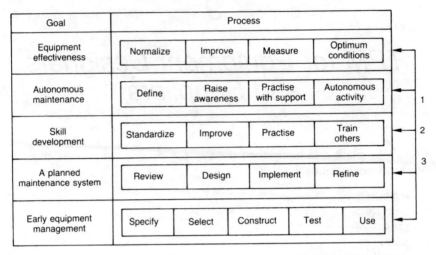

Parallel activities linked by:
1 design feedback
2 early warning systems
3 objective testing of new ideas

**Figure 9.1**  *Early equipment management: one of the five goals of TPM*

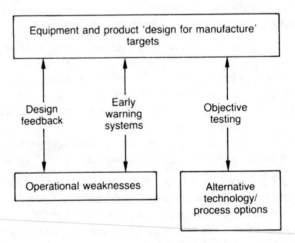

**Figure 9.2**  *Early equipment management: framework for maintenance prevention*

- counters
- checklists
- colour coding
- direction indicators.

Many other measures will result from development of the partnership.

Figure 9.5 illustrates the concept that two-thirds of the lifetime costs of new equipment is determined (but not spent) in the early design specification stages, and can therefore be said to be designed in. This serves to

**Figure 9.3** *Plant roll-out of TPM improvement process*

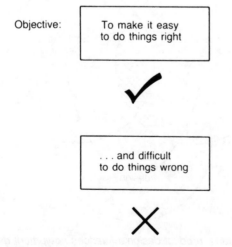

**Figure 9.4** *Improvements*

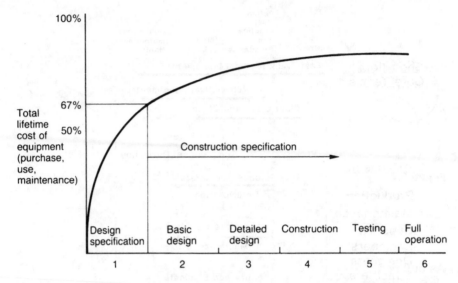

**Figure 9.5**  *Lifetime costing concept: two-thirds of our costs are fixed in the design stage (stage 1), so it must be right*

emphasize the importance of getting the design right first time, not just for intrinsic reliability but also for fitness for purpose, operability and maintainability.

The interrelationships between product design, equipment design and operation design are illustrated in Figures 9.6 and 9.7. There is a wide range of possible solutions, and the customer's requirements are crucial to the final choice.

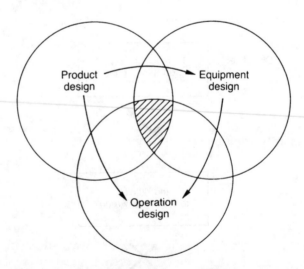

**Figure 9.6**  *Design issues: product design influences equipment design, and operations design is influenced by both*

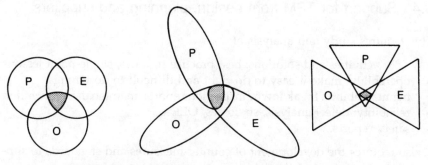

**Figure 9.7**  *Selecting the best design: many possible combinations of product, equipment and operation design. Customer requirements for timely, high-quality, low-cost products and services must provide the basis for selecting the preferred option.*

As outlined in the following sections, there are three major ways in which close collaboration between maintainers and operators and the five functional partners can ensure progress towards world class performance:

- objective testing
- design feedback
- TPM support.

## 9.2   Objective testing

This is technology/process design oriented and requires a search for new ideas using:

*Intrinsic reliability*   Repeatability of optimum conditions; simple construction; simple installation.
*Operational reliability*   Tolerance to conditions; simple manipulation; ease of maintenance.
*Lifetime costs*

All this is as part of continuous improvement.

## 9.3   Design feedback

This requires analysis of the product into components which are essential to its function and those which are required as a result of the design. As a general statement, the greater the value of the additional components compared with the essential components, the *lower* the design efficiency.

## 9.4 Support for TPM from design, planning and suppliers

This requires study and analysis of:

- TPM activities and solutions: best practice routines, single-point lessons
- operability: make it easy to do right and difficult to do wrong
- maintainability: breakdown/inspection reports, maintenance prevention
- reliability: defect analysis, six losses, OEE
- safety reports.

It also requires the development of counter-measures and standards to support the achievement of flawless operation in the areas of:

- maintenance
- breakdowns
- adjustments
- defects
- accidents
- dust and dirt.

Figure 9.8 illustrates the requirement placed on design engineers if they are to give full support to the TPM process. Figure 9.9 illustrates the approach necessary to define reasons for defects and ultimately to design out the weaknesses.

## 9.5 User-friendliness

The achievement of user-friendliness in equipment entails setting goals and determining measures which will progressively eliminate or simplify component parts:

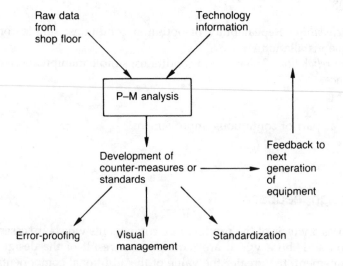

**Figure 9.8** *On-the-job routine for design engineers*

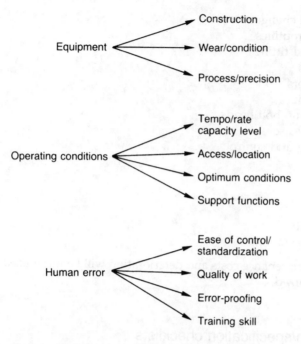

**Figure 9.9** *Early warning systems: defining reasons for defects using P-M analysis with the aim of designing out those defects*

- Can the part be eliminated?
- Can the part be integrated with an adjacent part?
- Can the part be a simpler shape?
- Are the strength, thickness, material appropriate?
- Are specifications for the part realistic?
- Can we standardize the part with other parts?

Other measures necessary to achieve user-friendliness are:

1  Collect feedback and use current data.
2  Analyse the part and ask 'why' five times:
  Consequences of failure?
  Causes of failure?
  Improve reliability?
  Improve maintainability?
  Set and maintain optimal conditions?

## 9.6  Standardization

Standardization is one of the essential approaches to achieving easy maintenance and trouble-free operation. Standardization can be applied to:

- operation procedures

- setups and changeovers
- asset care routines
- fixtures and fittings:
  adaptors
  connectors
  thread sizes
  screw, nut, bolt heads
  quick release
- monitoring and control:
  gauges
  oil
  heat
  electric
  pneumatic
  instrumentation.

As a result, maintenance costs and deterioration will be minimized and OEE will be maximized.

## 9.7   Design/specification checklists

These fall under three main headings.

### Effective design/specification

This concerns fitness for purpose, and is related to the earlier focus on user-friendliness.

- Can the item of equipment be eliminated? (Is it vital to the process or as a result of the design?)
- Can the item or part be integrated with the adjacent item?
- Can the item be simplified? (Can it be a standard part rather than a special?)
- Can we standardize the item with another item?
- Can the equipment/item cope with the environment? (dust/heat/dampness/vibration: adverse conditions as well as normal)
- Can the equipment control be simplified?
- Can the item be made of a cheaper/different material?
- Can a cheaper service be used?

### Operability

This is aimed at making it easy to do right, difficult to do wrong.

- Are frequent adjustments required?
- Are handles or knobs difficult to operate?

- Are any specialized skills or tools required for operation/adjustment? (startup, shutdown)
- Are blockages/stoppages likely? (How are they resolved?)
- Has any diagnostic function been built in? (glass panels, gauges, indicators)
- Startups and shutdowns: is additional manning required?
- How robust is the equipment? (Will the equipment break down or product quality be affected by poor operation?)
- Is the operator's working posture unhealthy?

### Maintainability

The keys here are to try to eliminate maintenance or to make it easy, infrequent and cheap.

- Can we eliminate the need for maintenance?
- Are areas easy to clean, lubricate or check?
- How long is the equipment setup time?
- How frequently does the equipment need tuning or calibrating?
- Are specialized maintenance skills required?
- Can failure be predicted?
- Have any self-diagnostic functions been built in? (Is it easy to find the cause of failures?)
- Can parts be easily replaced and plant restored quickly?
- How reliable is the equipment?
- Can we extend the maintenance interval?
- Does the equipment structure facilitate maintenance? (lifting heavy parts etc.)
- What routines are required?
- What spares support is required?
- Can breakdowns be restored cheaply? (Can spare materials and parts be purchased cheaply?)

Figure 9.10 shows how the seven partners can move towards world class.

## 9.8  Typical equipment design pilot project framework

The core team could be made up of:

- designer/specifier
- planner/specifier
- manufacturing engineer
- equipment operator
- equipment maintainer
- equipment supplier
- facilitator.

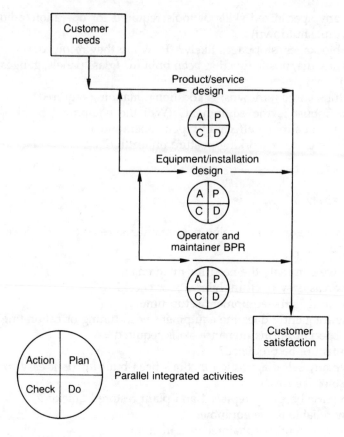

**Figure 9.10**   *Delivering world class performancre*

The key contacts consist of:

- purchasing
- finance
- quality
- product engineering
- process engineering.

A timetable for an equipment design project is shown in Figure 9.11. Initial training would involve the core team and the key contacts. The activity sessions are described in the following.

*Activity session exercises*

**Setting design targets**

- Criticality ratings.
- Design efficiency.
- Operational conditions.

| Days | 1 | 2 | 3 | 4 | 5 | 6 | 7 | 8 | 9 | 10 |
|---|---|---|---|---|---|---|---|---|---|---|
| Initial training day | ▫ | | | | | | | | | |
| Activity sessions | | ▨▨▨▨▨▨▨ | | | | | | | | |
| Presentation | | | | | | | | | ▫ | |
| Policy development and forward action plan | | | | | ▨▨▨ | | | | | |

**Figure 9.11** *Equipment design project timetable*

- Process trade-offs.
- Intrinsic reliability.
- Operational reliability.

It is highly desirable to involve the equipment supplier at this stage.

### Analysis, testing and assessment

- Confirm trade-off analysis of basic/outline on a modular basis.
- Establish testing/audit criteria for each module/subsystem.
- Conduct outline criticality assessment to predict equipment weaknesses per module.
- Establish criteria for standardization of components/spare parts.

### Refinement and simulation

- Confirm/refine each module/subsystem at a detail level.
- Incorporate errorproof devices for flawless operation.
- Establish tooling/maintainability criteria.
- Establish asset care regimes with supporting visual management.
- Simulate cleaning and inspection activities to improve operability.
- Simulate maintenance activities to improve maintainability.
- Feedback improvement suggestions.

### Construction

- Review construction constraints/opportunities.
- Agree quality audit milestones for main construction process.
- Define detailed project plan.
- Establish how the equipment will be located in relation to other equipment (layout considerations).
- Complete quality audit reviews.
- Establish best practice routines and develop training material.

### Trial/testing

- Project planning.

- Installation, including workplace organization and OEE measurement.
- First run trials.
- Confirm best practice and standardization.
- Joint sign-off of operation.

## Full operation

- Maintain normal conditions.
- Practise best practice routines.
- Strive to establish optimum conditions.
- Develop early warning systems and design feedback.

## Conclusions

Figure 9.12 shows how the nine-step TPM improvement plan may be used as part of the early warning system for asset care. Figure 9.13 shows how the improvement plan can be used in the design process.

Figure 9.14 shows a blank matrix for assessment of potential pilots. The value of alternative pilot projects can be assessed by scoring 1 or 3 in each of the six headings, i.e.:

- Does the project have known problems?
- Will improvement in the pilot impact on the plant/unit OEE?
- Can we measure the OEE of the pilot?
- What will be the visual impact in the area?
- Can we test the nine TPM implementation steps?
- Will it encourage teamwork between specifiers/users/suppliers?

Companies who adopt the philosophy of TPM for design will have the potential for a huge commercial advantage resulting from equipment with minimum total life cycle costs, which deliver high overall equipment effectiveness levels and flawless operation.

| | |
|---|---|
| OEE | Trend indicates need for action |
| Criticality assessment | Formal review of design performance post-installation |
| Condition appraisal | Audit/record of deterioration |
| Refurbishment plan | Record of life time costs |
| Asset care | Planned maintenance costs |
| Equipment history | Record of reliability |
| Six losses | Record of areas of improvement |
| Problem solving | Opportunities to pass on lessons learned |
| Best practice routine | Activities needed to achieve flawless operation |

**Figure 9.12**  *Using the improvement plan as part of the early warning system*

| | Design concept | Basic design | Detailed design | Build/install | Testing/ refine | Implement/ use |
|---|---|---|---|---|---|---|
| OEE | Set targets | Assess trade-offs | ✓ | ✓ | ✓ | ✓ |
| Criticality assessment | | ✓ | ✓ | | | ✓ Support training |
| Condition appraisal | Feedback on weaknesses | | | | Set standards | ✓ |
| Refurbishment plan | Assessment of lifetime costs | | | | | Monitor lifetime costs |
| Asset care | | ✓ | ✓ | | ✓ | ✓ |
| Equipment history | | | ✓ | ✓ | ✓ | ✓ |
| Six losses | Setting zero targets | | | | ✓ | ✓ |
| Problem solving | Target setting | | | ✓ | ✓ | ✓ |
| Best practice routines | | | | | Aim for flawless operation | ✓ |

**Figure 9.13** *Using the TPM improvement plan in the design process*

| Equipment pilots | Known problems | Impact on plant OEE | Ease of pilot OEE measure | Visual impact of pilot | TPM improvement plan | Teamwork | Total |
|---|---|---|---|---|---|---|---|
| | | | | | | | |

Rank: 1 = low 3 = high

**Figure 9.14** *Potential pilots*

# Case Studies

As emphasized in Chapter 1, application of the TPM process varies greatly according to the industry or enterprise where it is being applied. There is no better way of illustrating this divergence of approach than by presenting case study reports from the companies concerned.

Abbott Laboratories have found that TPM is complementary to and supportive of their target, which is to be one of the best manufacturing and distribution health care organizations in the world (Chapter 10).

Castle Cement has used the TPM process, but with its own interpretation as team production maintenance. This has been a key route to establishing and maintaining the best international standard of high-quality, efficient cement manufacture (Chapter 11).

The report from BP's Forties Alpha platform provides a fascinating illustration of the manner in which the TPM approach can be applied to offshore industry (Chapter 12). The concept is that they believe that TPM can and help to achieve an extension of the profitable life of the Forties Field, which is already one of the oldest in the North Sea!

Courtaulds Films Polypropylene provides an excellent example of where TPM has been used to secure earlier benefits of teamwork and integrated working as a means of generating significant extra capacity from existing equipment (Chapter 13).

Finally, Chapter 14 presents some brief cameo examples of the automotive industry's application of TPM.

# —— *10* ——

# *Abbott Laboratories*

*Kevin Smither*, Packaging Engineering Manager

## 10.1 Introduction

Abbott Laboratories Limited is the only mainland UK-based manufacturing and distribution affiliate of the Abbott International Corporation, a leading health care organization.

The UK manufacturing plant is situated in the south-east of England. It occupies a site of 116 acres and employs approximately 650 people. Its operations include chemical and pharmaceutical processing, packaging and distribution services.

In the manufacture of a wide variety of health care products, Abbott Laboratories believes in the following objectives:

- quality products without defects at all times
- a customer service level of 100%
- reduction in work in progress to one week of demand
- stable product costs.

To achieve these primary objectives requires commitment to:

- extraordinary ethical and safety standards
- involved, dedicated, well-trained work teams
- fast, simple and relevant systems.

As part of an overall strategy for improved performance, a number of complementary initiatives have been established. These include total productive maintenance (TPM), which offers the following:

- comprehensive quality
- high machine reliability
- planning and control
- maintenance prevention
- autonomous maintenance.

## 10.2 Why TPM?

During a seminar entitled 'Maintenance the Japanese Way', held at the National Motorcycle Museum, Professor Hajime Yamashina covered high machine reliability, comprehensive quality, autonomous maintenance planning and control, and many other valuable concepts. The issues considered through TPM would link well with other total quality techniques and would provide increased levels of efficiency and equipment effectiveness.

## 10.3   Steering group

A paper on TPM was presented to the senior management group, and as a consequence a steering group evolved to further investigate the possibilities for Abbott Laboratories. The steering group consisted of senior and middle management from both the engineering and the pharmaceutical production departments.

The steering group initially identified a liquid filling and packaging line for the TPM pilot, based upon the following:

- There were capacity problems.
- The product range was relatively simple.
- There were high maintenance costs.
- The operation exhibited variable performance.
- The operating staff were considered experienced and sufficient in number on which to build the foundations of TPM.
- The packaging line equipment was being relocated into a new improved area and thus it was considered an appropriate time to introduce change.

## 10.4   Scoping study

The steering group sought independent advice on the successful implementation of TPM. WCS International were invited to Abbott Laboratories.

The consultants conducted a scoping study and line survey and collected data and information under a number of headings, such as:

- current process organization structure for the plant
- flow chart of the main processes
- main/critical plant items
- amount and reasons for overtime
- working arrangements and practices
- recording and management of plant performance, quality and availability.

Interviews were conducted with plant operators, engineering technicians, and a cross-section of people representing other support groups, such as finance, quality assurance, planning, distribution and stock control.

The senior management group then reconvened for the presentation of results from the scoping study. Following this an announcement was made, and the TPM pilot was approved.

## 10.5   Getting started

In order to proceed it was necessary to assign TPM facilitators who were then trained over a two-day period. The facilitators included representatives from production operations and supervision, engineering and quality assurance, seven in total.

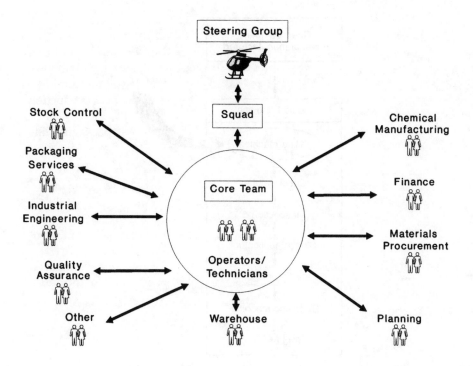

**Figure 10.1** *Abbott Laboratories: TPM framework*

Following facilitator training, a one-day introductory session was arranged for all those directly involved in the liquid filling and packaging operation together with a mixed squad representing all departments involved from manufacturing to despatch (Figure 10.1).

A core team was selected from the liquid filling crew and engineering technicians. A piece of equipment was specified (the liquid filling, plugging and capping machine), and a start date for the TPM pilot process was assigned.

## 10.6 Liquid filling process

The liquid filled is an inhalation anaesthesia product. It is manufactured at Abbott Laboratories in the chemical plant via a series of reaction and distillation processes.

The manufactured material is transported in bulk containers (Tycon or Byson vessels) into the pharmaceutical warehouse. It is then allocated to a work order and brought into the packaging area.

The bulk container is elevated and a series of connection pipes and intermediate holding tanks eventually result in it being coupled to the filling machine (Figure 10.2). Glass bottles are loaded by hand and are

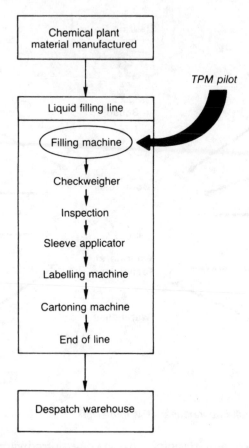

**Figure 10.2**   *Abbott Laboratories: product process*

volumetrically filled automatically. A plastic location collar is pushed on to the neck of the bottle followed by a cap which is torque tightened. The bottles are passed over a checkweigher and then undergo a visual inspection process before having a tamper-evident shrink sleeve applied. The bottles are labelled automatically, inspected, then cartoned or packed into air freight shippers as appropriate.

The liquid filling process from start to finish takes 2 minutes depending on size.

## 10.7   First core meeting

The core team of seven with the assistance of two main facilitators agreed to meet on a weekly basis for a period of twelve weeks.

The core team followed the TPM improvement plan for the filling machine. This included the following steps (Figure 10.3):

- machine criticality assessment
- machine condition appraisal

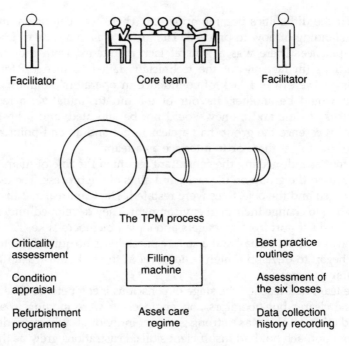

| Facilitator | Core team | Facilitator |

The TPM process

| Criticality<br>assessment | | Best practice<br>routines |
| Condition<br>appraisal | Filling<br>machine | Assessment of<br>the six losses |
| Refurbishment<br>programme | Asset care<br>regime | Data collection<br>history recording |

**Figure 10.3** *Abbott Laboratories: core team meeting*

- refurbishment programme
- asset care regime
- data collection and history recording
- assessment of the six losses
- best practice routines.

Throughout this process the consultants visited to monitor progress and support the team. The results of this investigatory work were then presented to the senior management group and the rest of the liquid filling line.

During the twelve-week period, through attention to detail and improving team performance, a number of small improvements were implemented at little or no cost. This had the effect of heightening the exposure of TPM and acting as a motivator for the team so that they continued to gnaw away at the longer-term issues and areas identified for improvement.

## 10.8 Team development

The formation of the core team consisted of appointed individuals, the reasoning being that as a new group it was important to balance the skills and characters in order that they would grow into a strong team.

During the first meeting the group established a start and finish point. They decided the format and set the target of the presentation of results to the senior operating board.

Then the difficulties became apparent. There was confusion and lack of understanding of how to proceed. The group was in a state of unconscious incompetence. There was a general lack of interest from the engineering technicians (in fact one of the technicians failed to turn up for the first meeting). There was a lack of confidence in operators' abilities; a concern that it would be another 'flavour of the month' initiative; a feeling that resources of time and money would not be allocated; and a host of other issues. In essence the group had a clear target and an end point but didn't know how to get there or to interface as a team.

At the second meeting the consultant assumed the role of main facilitator and focused the group on the tasks and aims of the process. The team issues submerged and the objectives were restated. The team realized they were in a position to change their own environment. They developed understanding and found support from managers and squad members alike. The emphasis was on attention to detail and a steady pace: this is no quick-fix solution. The team began to develop a high belief level in the task, but frustrations were beginning to creep in.

The team had reached the stage of conscious incompetence. The tasks had become clearer but operators and engineering were moving in an uncontrolled manner. This was a strong group of individuals who were developing a team approach but had insufficient skills. Frustrations grew as the need to control was recognized by the team. Members identified shortcomings in their colleagues, both personal and job related. This without question was the most difficult and painful stage in the team development. It required considerable help and support and was the longest phase in the team's progression.

Throughout the twelve-week period other industrial relations problems arose. The rest of the line crew were not being informed of the team's activities, felt isolated, and offered less and less support. The tensions between those in the team and those not, ballooned at an alarming rate. The resolution to this problem was fairly quickly found. Structured feedback to the crew was initiated through a weekly half-hour line meeting, together with a TPM log which was kept on the line for reference.

During this phase it was often necessary to restate the objectives, help the team identify key recommendations, and act quickly and effectively when solutions to problems were found. At this point it was important for the team's morale and self-esteem to see recommendations carried out.

## 10.9   Team problem solving

Throughout the pilot process a number of low-cost and no-cost improvements were identified. This was due to fine attention to detail and cooperation between operators and team members alike. Since there had been minor hassles both identified and hidden that no one had time to tackle, they had remained. For example:

- The locating collar feed bowl would sporadically jam up, stopping production. With time a technician improved the performance by a factor of 4:1 on the exit from the vibratory bowl, and by 14:1 on jam-ups inside the bowl.
- When changing from one product to another, the pumps and filters took an hour or more to prime. Investigation ensued and new filters were fitted that required no priming, and this improved downtime.
- The bottle infeed sensor was replaced from an electromechanical switch to an electronic device with no moving parts: this became more reliable.

A high-speed video camera was made available to aid diagnostic work.

It came to the attention of the team during the study of best practice routines that the twilight crew (employees working 6.00 p.m. until 10.00 p.m.) had a better method of operation. This was written into the basic operational procedures manual and is now standard.

A programme of refresher training for all operators was initiated based upon the reviewed best practice routines. Consequently all tasks are now performed the same way.

The team were able to identify very quickly problems, hassles and minor defects, but they were fairly inexperienced at finding practical solutions for all problems. An example of this is when access to part of the machine for weight/fill control adjustment was hampered by an intermediate bulk container bolted to a pallet. This was found to be awkward to work round. Castors were purchased (low cost) and fitted, the pallet was dispensed with, access was made simple, and the intermediate bulk container was manoeuvrable. However, the bulk container when full was top heavy and, having lost the wide pallet for a base, was tipped over, causing loss of product and delays. The container was fitted with additional legs and castors and the problem was eventually resolved.

This highlighted the need for tighter control on problem solving techniques. Team training was arranged covering a six-step model:

1  Identify the problem.
2  Brainstorm causes.
3  Analyse data.
4  Brainstorm solutions.
5  Reach consensus.
6  Develop/implement action plan.

This proved most beneficial (Figure 10.4).

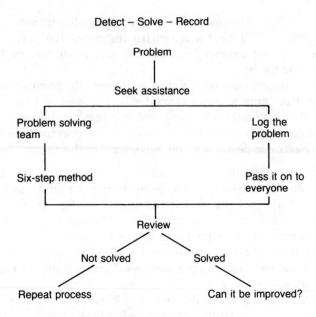

Detect – Solve – Record

**Figure 10.4**   *Abbott Laboratories: team problem solving*

## 10.10   TPM process

The core team then began planning the roll-out to the rest of the line, to include all other pieces of equipment and all other line operators. The introduction of other engineering technicians was also planned at this stage.

The core team had developed a level of confidence and enthusiasm that resulted in a very tight schedule. It was considered the best option to allow them to proceed.

## 10.11   Conclusions

There are now (early 1993) four regular weekly meetings, one with the main core team and three subgroup meetings. Each subgroup comprises some core team members and newcomers (Figure 10.5).

Each subgroup will develop the TPM process for an assigned piece of equipment (shrink sleeve applicator, labelling machine, cartoner etc.). Then they will produce an action plan that will be augmented into an overall scheme of continuous process and procedural improvements in order to achieve world class manufacturing status. They are approximately half-way through this exercise (see Figure 10.6).

As the team develops it will require reduced facilitating and should be able to work autonomously. Initially another critical line will be adopted, followed by others as more experts are groomed.

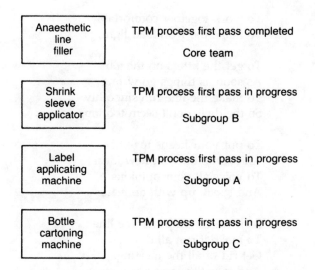

| | |
|---|---|
| Anaesthetic line filler | TPM process first pass completed<br><br>Core team |
| Shrink sleeve applicator | TPM process first pass in progress<br><br>Subgroup B |
| Label applicating machine | TPM process first pass in progress<br><br>Subgroup A |
| Bottle cartoning machine | TPM process first pass in progress<br><br>Subgroup C |

**Figure 10.5**  *Abbott Laboratories: interim stage in TPM implementation*

| | |
|---|---|
| Seminar 'Maintenance the Japanese way' | April 1991 |
| Presentation to senior management | August 1991 |
| Scoping study presentation of results | Feb./Mar. 1992 |
| TPM training and awareness sessions | July 1992 |
| TPM pilot start date | August 1992 |
| Pilot complete: presentation of results | October 1992 |
| Roll-out plan | December 1992 |
| Implement roll-out | January 1993 |

**Figure 10.6**  *Abbott Laboratories: path to TPM implementation*

Abbott Laboratories have found that TPM is complementary to and supportive of our target to be one of the best manufacturing and distribution health care organizations in the world. As we found:

*A team member's interpretation of what TPM means*

> The Abbott's improvement scheme
> Is what it's all about
> Improvement of machinery
> To get the product out.
>
> To use your own initiative
> To get the system right

To work together comfortably
To make the workload light.

To get the fitters on the job
As soon as things go wrong
To make the line run smoothly
So the days don't seem too long.

To put your heads together
And come up with new ideas
To voice all your opinions
And speak up with no fears.

So if we can improve the line
To make it run all day
Get rid of all the meetings
And take the cares away.

# — 11 —

# Castle Cement

*Roger Olney*, Engineering Manager

## 11.1 Background

Castle Cement's Ketton manufacturing plant consists of a two-kiln line plant, one kiln producing 3000 tonnes per day and one kiln producing 1000 tonnes per day.

The raw materials are quarried and ground on site (Figure 11.1). The clinker produced is also ground on site and converted into product which leaves the plant mainly by bulk road transport. The fuel used in the kilns is coal which is received by road and ground on site to a powder. The only other material imported on to the plant is gypsum.

The plant is relatively modern. The largest kiln is fully computer controlled and one of the most modern in Europe.

## 11.2 Maintenance methods

Maintenance is carried out utilizing a fully integrated computerized maintenance management system (Fluor Daniels CMMS). The system has approximately 40 screens installed on site, which includes the workshop, foremen, stores, accounts etc.

Maintenance consists of major shutdowns planned in detail for the two kilns with regular PM on auxiliary equipment that can be stopped outside the kiln repairs. The major shutdowns are planned using a modern critical path analysis program (OPENPLAN) to reduce shutdown lengths to the minimum.

Significant condition monitoring is carried out using oil analysis and vibration analysis equipment. Full oil analysis is carried out on major drives off site, and there is on-site analysis to a lesser degree on other drives. In particular we use a PQ analyser as an early indication for both major and minor drives. Ultrasonic, magnetic particle and boroscope inspection techniques are used together with flow and pressure measurement monitoring. Data are analysed by PC, taking into account oil and vibration results for impending problems.

The maintenance craftsmen are multi-skilled, as are the process operators.

## 11.3 Need for improvement

Better teamworking between maintenance and production was needed to solve ongoing plant problems which were not necessarily related to

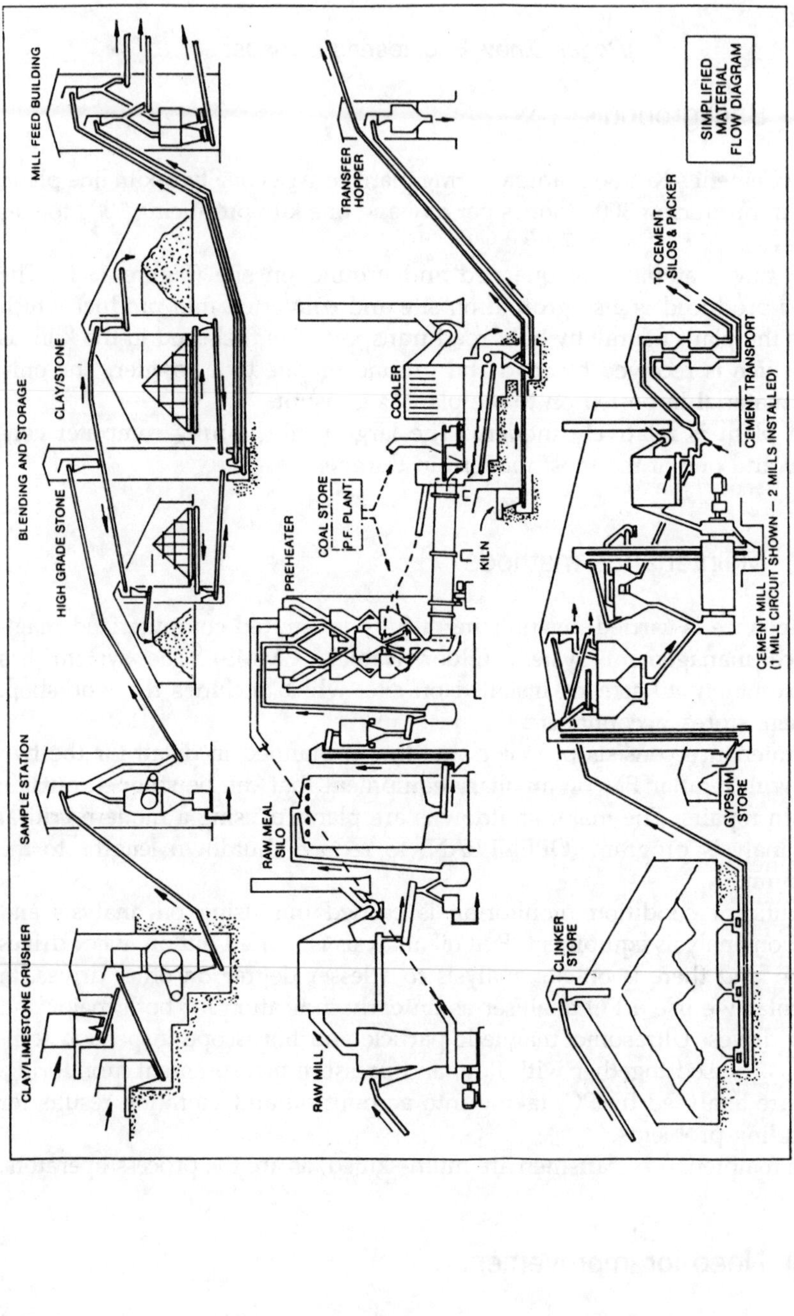

**Figure 11.1**   *Castle Cement: simplified material flow diagram*

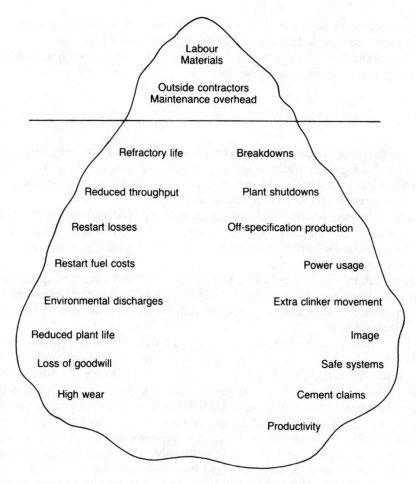

**Figure 11.2**  *Castle Cement: visible and hidden maintenance costs*

machinery breakdown (Figure 11.2). Most of the major problems had been eliminated but run factors and production rate needed to be improved consistently.

Better ownership of plant by both maintenance and process operators was required. Process operators for example are totally flexible across the plant without responsibility for equipment.

The maintenance strategy betterment policy includes monitoring of current methods. The likely benefit of proactive maintenance is currently being looked at.

## 11.4   Initial sell

The first introduction to TPM was at an optimum maintenance seminar in June 1989 when I thought the methods employed looked advantageous.

Senior engineers were introduced to TPM at the 1991 maintenance seminar and their response further convinced me that TPM could help us.

Peter Willmott was approached to carry out a presentation to senior management and supervision on site.

## 11.5   Justification

Before the presentation I needed to find ways to justify the training expense, and a suitable project area to assess any improvement made. The project area ideally needed to have scope for improving the run factor, to be accessible for repair work, and to have a control to compare.

The only area which fitted the bill completely was part of our cement milling line. The building was on its own, had two identical plant lines, and had room for improving the run factor, although this was mainly due to over-capacity in the area.

Justification was difficult owing to the almost 30% in-built over-capacity of the lines. Downtime costs were low unless production could not be reached, which was unusual. Scrap or off-specification product was rare, or at least was thought to be.

The power consumption of the equipment was relatively high (around 7 MW) per line, and with the large difference between day and night rates some additional night running might be justifiable. Costing night failures by the difference between day and night tariffs suggested that there was a possible annual saving of around £200 000 if these night stoppages could be avoided. This saving would be as a result of an estimated 12% improvement in OEE. Thus the value of each £% improvement in OEE would be £17 000 for numbers 9 and 10 cement mills only. If the spare capacity could be sold, the commercial benefits would be significantly higher.

Armed with sufficient justification to cover a scoping study and training, I obtained permission for a presentation. Following successful presentations to supervisors and senior managers, I sought and was granted permission to carry out a scoping study and a pilot scheme.

## 11.6   Plant sell

A notice to the plant was circulated informing of a study into a pilot scheme for the chosen area. Following the scoping study which included a number of interviews with staff, a proposal was drawn up for a pilot scheme in May 1992. Prior to the pilot scheme commencing, some facilitator and key contact training was given, comprising a TPM process compressed into four days.

The facilitator training was followed by squad and team training together with plant awareness sessions to some 200 plant personnel. For these sessions, TPM was defined as team production maintenance to help acceptance and break down barriers (Figures 11.3, 11.4). The projects worked on during these training sessions revealed unexpected potential savings of around £60 000.

*Vision*
Achieving world class status through cost effective manufacturing and excellence in all that it does

*Based on goals:*
- High level of reliability
- High quality levels

*Achievement with commitment to:*
- Develop capable, motivated people
- Working as teams
- Dedicated to achieve high levels of equipment and process capability

TPM is a cornerstone to achieve the vision, goals and commitment

**Figure 11.3** *Castle Cement: the vision*

CASTLE
CEMENT

Team production maintenance

*Team production maintenance is being introduced at Ketton as a key route to establishing and maintaining the best international standard of high-quality, efficient cement manufacturing. It will use the capabilities of all employees, recognizing that the knowledge and commitment of all, working as a team for a common purpose, will enhance the competitive position of the organization.*

**Figure 11.4** *Castle Cement: team production maintenance*

## 11.7  Pilot scheme

The pilot scheme started properly in mid-July with four projects.

The strategy for the scheme effectively was to take four multi-disciplinary teams through the TPM process on four separate projects over a six-month period, the bulk of the work being carried out in a three-month period (Figure 11.5). The teams consisted of a team leader, an operator, two day craftsmen, a shift craftsman and one other. The team was backed up by the squad, of which 15 or so had been through a compressed TPM training exercise. The individual shifts were varied to allow the team to meet weekly to review results and agree action.

The projects were chosen from areas where improvements were known to be possible:

- weigh feed conveyor
- lubrication system
- Poldens transport system
- clinker extraction.

Other parallel projects were:

- general spillage projects
- dust extraction resolution
- pest control resolution
- fitting shop CAN-DO
- site CAN-DO.

| Milestone | June | July | August | September | Oct. – Jan. |
|---|---|---|---|---|---|
| 1 | TPM policy statement | | | | |
| 2 | Agree pilot 1 | Set up project structure | | | |
| | | Clarify PMT role and implement on pilot 1 | | | |
| 3 | | Appoint facilitator and teams | | | |
| 4 | Outline OEE measures | Set targets and STDS | Begin measurement process and | | |
| | Do RP calc. | | | refine | |
| 5 | Begin awareness cascade | Start detailed TPM training for pilot 1 | | | |
| 6 | | Set up condition appraisal for pilot | | | |
| 7 | Do CA | Link to new ways of working | | Begin PM process | |
| 8 | | Clarify systems support using pilot as template | | | |
| 9 | | | | Achieve step 4 of AM | |
| 10 | | | | | Pilots 2 to 4 |
| 11 | | | | | Continue PM on pilot 1 + Continuous improvement |
| 12 | | | | | Start step 3 above for pilots 2 to 4 |

**Figure 11.5**  *Castle Cement: TPM pilot programme*

The teams went through the following basic steps:

1 List and describe assets.
2 Ask key managers and supervisors to assess criticality independently from their perspective.
3 Collectively weigh and rank.
4 Facilitator and team leaders ask operators and maintenance staff (team members) for a condition appraisal.
5 Collect view on:
   • refurbishment needs
   • checklists, inspections
   • planned, preventive schedules
   • spares requirements
   • support equipment
   • standard operating procedures
   • skills assessment and training requirements
   • potential for chronic problem resolution.

Some examples of this process follow the steps taken from one of our four project presentations (the Poldens unit in number 9 cement mill). Included are:

• criticality assessment (Figure 11.6)
• condition appraisal and refurbishment (Figure 11.7)
• future asset care (Figure 11.8)
• OEE calculations and financial assessment (Figures 11.9–11.12)
• example of results after improvements (Figure 11.13).

| Equipment description | 1–3 ranking as impact on: | | | | | | | | | |
|---|---|---|---|---|---|---|---|---|---|---|
| | EOR | R | PQ | MI | TPV | LOP | S | ENV | C | Total |
| Compressor | N/A | 1 | 3 | 1 | 3 | 1 | 1 | 1 | 3 | 14 |
| Pipelines | 3 | 3 | 3 | 3 | 3 | 3 | 1 | 3 | 3 | 25 |
| Dust plant | 3 | 3 | 1 | 3 | 1 | 1 | 1 | 3 | 3 | 19 |
| Top tank | 1 | 1 | 3 | 1 | 3 | 3 | 1 | 1 | 3 | 17 |
| Pressure vessel | 3 | 3 | 3 | 3 | 3 | 3 | 1 | 1 | 3 | 23 |
| Electrical | 1 | 1 | 3 | 3 | 3 | 3 | 1 | 1 | 1 | 17 |

| | | 1 | 3 |
|---|---|---|---|
| EOR | = ease of repair | easy | difficult |
| R | = reliability | high | low |
| PQ | = product quality | low | high |
| MI | = maintenance interval | long | short |
| TPV | = throughput velocity | low | high |
| LOP | = loss of production | low | high |
| S | = safety | low | high |
| ENV | = environmental | low | high |
| C | = cost | low | high |

**Figure 11.6** *Castle Cement: criticality assessment*

| Condition appraisal report | | | | Person days | | |
|---|---|---|---|---|---|---|
| Plant no. Description | Condition appraisal | Refurbishment required | Cost | E | M | Op |
| 12.04 Compressor | Cooling water | Improved filters | High | 1 | | |
| | Area untidy | Better housekeeping | Low | | | $\frac{1}{4}$ |
| 12.03 Receivers | Area untidy | Improve housekeeping | | | | |
| | Poor valve identification | Labels required | Low | | $\frac{1}{2}$ | |
| Pipework | Poor valve identification | Labels | Low | | $\frac{1}{2}$ | |
| (motor room) | Gauges useless | Replace | High | | $\frac{1}{2}$ | |
| | Aeration pipes blocked | Non-return valves | High | | 1 | |
| | Water contamination | Extra moisture traps | High | | 1 | |
| (Poldens pit) | Misalignment of conveying pipework | Realign | Low | | 2 | |
| | Access to pipe flanges | Improve ground clearance | Low | | 1 | |

**Figure 11.7**   *Castle Cement: condition appraisal*

The results of this project were and still are very good, and the predicted savings of £43 000 should be realized. The results from this and the other three projects were presented in October 1992.

The potential savings produced from these projects were of the order of £200 000. Savings from areas not previously identified were given in the reports.

## 11.8   Payback

One project alone has paid for all training to date and any planned in the future as regards TPM.

The results of a team of people spending several weeks thinking about and studying a problem during their normal working is really very illuminating. The standard of reports is very good indeed.

The results in terms of downtime have been demonstrated for one of the projects. Total project savings will approach £400 000 (Figure 11.14).

Payback in terms of teamworking is more difficult to measure in an isolated and relatively small project. The general attitude to housekeeping and ownership has been mostly enthusiastic.

## 11.9   Problems

There is a problem of acceptance at the supervisor level. At the grass roots, acceptance is mostly enthusiastic. One of the reasons is that fitters and

| | Responsible |
|---|---|
| *Daily routine (per shift) checks* | |
| ● Check entire system for leaks | Operator |
| ● Check instrument air pressure (6 bar) | Operator |
| ● Check control lamps | Operator |
| ● Check drains | Operator |
| ● Check compressor, alarms, noise, etc. | Operator |
| ● Ensure area is clean | Operator |
| | |
| *Daily routine (24 hours) PM* | |
| ● Check air lubs | Operator |
| ● Check pulse air pressure on dust plant | Operator |
| ● Check emission from dust plant stack | Operator |
| ● Check dust plant fan for vibration: clean fan if necessary | Operator |
| ● Check operation compressor/house ventilation plant | Operator |
| ● Ensure area is clean | Operator |
| | |
| *Daily routine (night shift)* | |
| ● Polden trends | Operator |
| | |
| *Weekly routine* | |
| ● Grease dust plant fan | Operator |
| ● Record Polden blows | Operator |
| | |
| Weekly routine | |
| ● Compressor checks | Maintenance |
| | |
| *Fortnightly pre-routine checks* | |
| ● Check mushroom and slide valve | Maintenance/shift |
| ● Check outlet valve | Maintenance/shift |
| ● Check vent lines | Maintenance/shift |
| ● Check system for leaks | Maintenance/shift |
| ● Check pipe thickness | Planning |
| | |
| *Fortnightly routine* | |
| ● Inspect 3 in vent line (blockage) | Maintenance |
| ● Check operation 3 in vent valve | Maintenance |
| ● Inspect conveying line for blockage | Maintenance |
| ● Access through 4 in non-return valve | Maintenance |
| ● Check annular rin (pressure vessel) and 6 in aeration nozzles | Maintenance |
| ● Check two aeration nozzles, buffer hopper | Maintenance |
| ● Inspect pressure vessel inside for build-up of cement | Maintenance |
| | |
| *4-weekly routine* | |
| ● As above | Maintenance |
| ● Check 6 in vent valve | Maintenance |
| | |
| *26-weekly routine* | |
| ● All above tasks | Maintenance |

**Figure 11.8**  *Castle Cement: future asset care*

| Week no. | Poldens run time (hours) | Mill run time (hours) | Mill output (tonnes) | $\dfrac{\text{Poldens tph}}{\text{Mill tph}} \times 100\%$ | Cost associated with Poldens inefficiency (£) |
|---|---|---|---|---|---|
| 28 | 138.5 | 120.5 | 12 245 | 87.0 | 271 |
| 29 | 151.8 | 138.25 | 14 110 | 90.9 | 204 |
| 30 | 144.7 | 118.25 | 11 068 | 81.8 | 399 |
| 31 | 148.3 | 135.5 | 13 669 | 91.4 | 193 |
| | | | | Average 87.8% | Total £1067 |

**Figure 11.9** *Castle Cement TPM project (Poldens unit in number 9 cement mill): performance rate analysis*

| Week no. | Lost time due to Poldens (hours) | Mill run time (hours) | Availability (%) | Cost of lost production time (£) |
|---|---|---|---|---|
| 28 | 16 | 120.5 | 88.3 | 1754 |
| 29 | 0.75 | 128.25 | 99.5 | 82 |
| 30 | 12 | 118.25 | 90.8 | 1316 |
| 31 | 1.5 | 135.5 | 98.9 | 164 |
| | | | Average 94.4% | Total £3316 |

**Figure 11.10** *Castle Cement TPM project: availability rate analysis*

| Week no. | OEE (%) | Availability (%) | Performance (%) | Quality (%) |
|---|---|---|---|---|
| 28 | 76.8 | 88.3 | 87.0 | 100 |
| 29 | 90.4 | 99.5* | 90.9 | 100 |
| 30 | 74.3 | 90.8 | 81.8 | 100 |
| 31 | 90.4 | 98.9 | 91.4* | 100* |
| Average | 83.0 | | | |
| Best of best | 90.9* | 99.5 | 91.4 | 100 |

**Figure 11.11** *Castle Cement TPM project: OEE calculations*

| | Performance | Availability | Total |
|---|---|---|---|
| £/% per month | 87.5 | 592 | |
| £/% per annum | 1 137 | 7 696 | |
| Achievable improvement (%) | 3.6 | 5.1 | |
| Savings/annum (£) | 4 093 | 39 250 | £43 343 |

**Figure 11.12** *Castle Cement TPM project: financial assessment*

*Performance rate analysis*

| Week no. | Poldens run time (hours) | Mill run time (hours) | Mill output (tonnes) | $\dfrac{\text{Poldens tph}}{\text{Mill tph}} \times 100\%$ |
|---|---|---|---|---|
| 42 | 137.85 | 126.26 | 12 334 | 91.6 |

*Availability rate analysis*

| Week no. | Lost time due to Poldens (hours) | Mill run time (hours) | Availability (%) |
|---|---|---|---|
| 42 | 1.5 | 126.26 | 98.8 |

*OEE calculations*
*For week 42*    OEE $= 91.6 \times 98.8 = 90.5\%$
*For weeks 28–31*    OEE (average) $= 83.0\%$
                (BOB) $= 90.9\%$

**Figure 11.13** *Castle Cement TPM project: effect of improving bottom vessel aeration*

operators are identifying problems not previously recognized by the supervisors. The fitters and operators who are closer to the equipment are recommending PM to the supervisors, who would previously have determined this themselves.

Shift pattern differences caused considerable problems with meetings of the teams, and led to dissolution of the teams following the project presentations.

| Project potential savings identified | |
|---|---:|
| L/C tunnel belt | 33 800 |
| 7RM Atox rejects | 20 880 |
| Skako feeders | 23 500 |
| Lubrication units | 92 000 |
| Lubrication units capex | 100 000 |
| Poldens | 86 000 |
| 8 coal mill | varies |
| RM09 | 13 000 |
| 9 and 10 weighfeeder control | 54 000 |
| 8CM elevators | 5 000 |
| | £378 000 |
| To date conservative estimate | £150 000 |

**Figure 11.14**   *Castle Cement: payback*

## 11.10   Roll-out

A policy statement has been developed for TPM over a three-year pro-gramme. The statement incorporates all aspects of the current maintenance strategy including OEE targets for major plant. Projects and training have been set up and training has commenced. The strategy is as follows:

*Implement pilot recommendations*
- Meeting between supervisors and team leaders/representatives.

*Establish local TPM champions*
- To decide priorities.
- To establish teams including operators, shift maintenance and day main-tenance.
- May be existing teams.

*Establish a TPM infrastructure*
- Allocate squad by production area.
- Set up single-point training plans, initially TPM improvement plan then skill training.
- Set expectation of minimum time for TPM improvement time: ideally allocate a part of the week for TPM.

The roll-out actions are as follows:

*Build up enthusiasm for TPM*
- TPM pilot feedback session for site.

*Local TPM champion training*
- TPM improvement plan.
- Teamworking.
- Problem solving.
- Visual management.

*Set local improvement targets*
- Improvement to best of the best.
- Improving the quality of asset care (zero breakdowns).

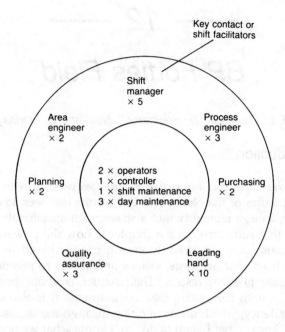

**Figure 11.15** *Castle Cement: roll-out team and squad*

Five teams are currently (1993) involved in new projects, with a further ten projects due later (Figure 11.15). The new projects effectively use training on the job as those previously, but the emphasis is on on-site training rather than external training. The facilitator, who was trained in the initial sessions, was appointed full-time recently as being the guiding light.

It is hoped that any plant improvements or repairs will be completed by the teams and that the teams will stay together for the completion of the projects. The aim of this training was to train 45 shift operators and 45 day maintainers. This process will have reached full implementation for 15 pieces of equipment and 30 under improvement/implementation by December 1993.

In parallel with this, work is under way to fully implement the pilot projects, a good deal of which has been carried out to date.

## 11.11 Conclusions

A number of unexpected bonuses have been apparent during the exercise to date. The potential return in terms of profit is substantial, and the project has been self-financing. It has resulted in better relationships between maintenance and production, with better ownership of plant.

We did not expect to achieve results overnight, but the potential is very significant and we are committed to the longer-term plan.

# —— 12 ——

# BP Forties Field

R. L. Thomson, B. Howden, T. Neill and S. Lindsay

## 12.1 Introduction

This chapter describes the innovative techniques used to rejuvenate one of
the grand old ladies of the North Sea. The results has been to change what
was becoming a siege mentality into a strategy for sustainable profit.

The thread that runs through the chapter is how the powerful impact of
empowered teams has produced successful change based on results. The
themes range from abstract ideas, visions and values to practical processes
that demonstrate positive results. This chapter is about people and the
success that lies with harnessing their commitment. It is also about achieving interdependency, which means moving through the stages of 'I am told
what to do,' 'I know what I want to do,' to 'I know what we need to do to be
the best at what we do.'

This chapter, written by technicians and supervisors at the sharp end, is a
demonstration in itself of empowerment in action.

## 12.2 Empowerment

Today's businesses often have to operate in an environment of constant and
sometimes dramatic change. Change is driven by the chaos of competition,
by technological advance, and the vagaries of the domestic economy and
increasingly the global market-place. More and more organizations are
focusing their efforts on emphasizing continuous process improvement,
and on the service of increased return to customers, shareholders and also
employees. The prime prerequisite for these objectives is valuing, developing
and fully utilizing the skills and the competencies of the whole workforce.

It has now become clear that, in the future, companies are unlikely to be
able to offer lifetime security of employment. This is reflected in the commitment employees are prepared to give. For the company, the return on human
investment is a direct function of the value placed upon people. For employees, autonomy, increased responsibility, involvement in the decision process,
challenge, skills growth and personal marketability, as well as financial
reward, are the new values.

Leaders in world class organizations are now striving to build and maintain structurally flat, high-involvement cultures. These organizations provide autonomy and allow real decision making power, involvement and
ownership to be devolved to the operational front line. In future, decisions
at all levels will comprise elements of creativity, innovation, calculated risk

and, above all, belief in possibilities. This unleashes previously hidden talent and creates an environment, within our organizations, that can emulate the flexibility, responsiveness and fleetness of foot sometimes more prevalent in our smaller counterparts.

A culture change process does not happen by itself: it needs a helping hand. Forties employed the services of consultants in order to clarify and provide support for that drive. We started by identifying the critical success factors. These serve as a catalyst and a reason for change and also provide a focus for our change efforts. After that came our mission statement: 'To extend the Forties Field life to 2010 by being a role model for operating a declining oilfield.' Easy to say; an enormous challenge to achieve. The visions and values are the building blocks of the culture change and underpin the new direction for Forties into the future. We sought to become interdependent and to be capable of building rich, enduring and highly productive relationships with colleagues and other companies. Out of this evolves sustained business improvement. To be successful with others we must pay the price and first make a success of ourselves. To quote Samuel Johnson: 'There can be no friendship without confidence and no confidence without integrity.'

The 'how' to reach the goal is a function of achieving excellent business results, which, in turn, reflect the values of the organization, the attitudes, mind-sets and beliefs that determine how employees interact with each other and their customers: the teamwork, the collaboration, the involvement, the speed of response, the trust and that new phrase that summarizes it all – the empowerment of the people.

Empowerment for us means creating a sense of job ownership by giving clear direction, having control of resources, real responsibility and appropriate coaching; it means offering help without removing responsibility. In practice that necessitates leaders establishing clear goals and expectations, as well as providing the tools to complete the task. It means encouragement, trust, integrity, recognition that our associates are an integral part of our day-to-day lives and listening and responding to their and others' needs, with respect and with empathy.

As with all good things, the more you put in, the more you get out. The foundation for success is a clear determination and total commitment to see the process through, from both the top and the bottom of the organization. Half measures are a folly. That way lies the route to a raised anticipation for a new way forward, which, if not driven firmly but fairly to fruition, leads to unfulfilled expectations and therefore disappointing bottom line deliverables.

Throughout Forties, programmes were instituted to raise empowerment awareness and convey to the staff population the key principles, concepts and visions and values inherent in the process. Many Forties personnel, from manager to operations technician, were trained as trainers and facilitators. It was interesting and encouraging to see the Forties Asset Team Leader, being coached by a production technician.

We organized ourselves into individual platform assets and broke that down into operational teams. We then set about applying our new knowledge with a view to pushing out the boundaries at our worksite and testing ourselves in what for us were previously uncharted waters.

To say that there was no resistance to change, and also to convey empowerment as some kind of pseudo-religious experience, is of course not correct. In certain sections and among some employees, corporate inertia and resistance to change, any change, were there as always. Several syndromes manifested themselves:

> It's the boss's new flavour of the month. We'll pay lip service for a while and perhaps both he and this latest fad will disappear.

After a while:

> God! This is serious. (to the accompanying sound of *some* heels digging into the concrete)

Eventually:

> This is only applied common sense and anyway, it's nice having a bit of a say for a change. It can even be fun.

Recession, downsizing, job insecurity are not a fertile ground to breed the psychological attitude needed to facilitate a change process. People naturally hang on to their comfort zones, especially if their part of the organization had been run on strictly autocratic and authoritarian lines. Also, how difficult is it for the manager, who, accustomed to direct accountabilities, takes the lid off the pressure cooker whilst still trying to accurately focus the direction of the steam? This is not about setting people free and breeding a kind of corporate anarchy. It is about allowing employees the space to be all that they can be and to harness the ideas and enthusiasm that come from being a valued member of the team. For the manager, this means developing a whole new armoury of personal communication and person management skills, not least of which is to understand that operatives at the sharp end really can be trusted to deliver extraordinary returns. All that is required is to delineate the task, give the trust and most of the time one will be astounded by the results.

However, it takes an act of faith to jump over the wall. Who knows how hard the landing might be. Initially, it is much harder to operate in this manner. After a while, the bottom line deliverables begin to fall out of the culture change process.

Eventually for most of us came the realization that if we did not jump then circumstances would push us over. We simply had to do more with less. Within this environment we would at least have the tools to navigate down towards 2010 in a controlled manner and choose the landing spot.

To say that we have reached adolescence is about right. There is a realization that to produce oil longer, we have to do it smarter. To achieve that, we must use all the weapons in our arsenal. Utilizing these hard-won skills, we intend to reach the goal that we have set ourselves.

Within Forties we now have empowerment concepts and the solid foundation of culture change. The vision and values stem from the Forties mission statement, which is: 'We shall extend the field life to 2010 by being a role model for operating declining oil fields.'

Achieving the mission statement is based on:

*Integrity*   Openness, trust, high ethical standards and respect for others in dealing with any individual or organization.

*Teamwork*   Individuals working cohesively with a common sense of purpose to achieve business objectives.

*Empowerment*   A supportive environment in which people are given both the authority and the resources to make sound decisions within established boundaries.

*Knowledge and skills*   Recognizing, valuing and developing the knowledge and skills of our people as a vital resource.

*Ownership*   Willingness to get involved and take a personal responsibility for meeting the Forties challenge.

In the move towards a more empowered climate, the day-to-day behaviour used by supervisors and managers is crucial to Forties successfully achieving its goals.

A number of themes are central to this empowering role:

- letting go
- less direct control
- developing job ownership
- developing skill and knowledge in others
- thinking longer term
- providing direction
- providing support
- adopting a wider role.

## Letting go

This is at the very heart of empowerment and, for many people, is the hardest thing to do. This means pushing decisions as far down the organization as possible, ensuring that authority and responsibility are clearly understood and accepted.

## Less direct control

Empowered teams take more control of their jobs and processes – monitoring quality and output. Managers and supervisors need to stand back from the details. This does not mean throwing caution to the wind. It means trusting a properly skilled and resourced workforce.

## Developing job ownership

This applies to managers and supervisors as well as their teams. It means having more self-direction and not constantly passing issues upwards. Job ownership comes from teams and individuals having meaningful challenging work and being involved in how to meet agreed goals.

### Developing others' skill and knowledge

Without this, the previous themes cannot be met. The manager and supervisor have a key role in identifying opportunities and developing the talents in their teams to the full. This means less telling and more coaching and training.

### Thinking longer term

More reliance on the team releases more time for the manager and the supervisor. Proactivity becomes the order of the day, with less fire fighting and more fire prevention. Thinking becomes strategic and longer term, a subject previously left to the boss.

### Providing direction

As teams take on more responsibility for their jobs and processes – how work is done – they need to be clear on what is expected of them. Managers and supervisors need to provide information on business goals and ensure that effort and skill is oriented in the right direction.

### Providing support

Providing support without removing responsibility is one of the keys to a motivated, empowered workforce. Support comes in many forms:

- ensuring resources are available
- encouraging ideas and initiative
- coaching
- recognition
- feedback on performance
- providing information.

### Adopting a wider role

As teams broaden and expand their roles, so do managers and supervisors. They gain more knowledge and develop more skills. They may lead more teams and fully realize their own potential. They do work previously carried out by their boss and are less protective of their own turf. Managers and supervisors are the agents of changes.

People adapt in various ways to the changing role. It comes easier to some than others. Being an empowering leader requires skills, the successful application of which takes time and practice. Together we are building an environment where these behaviours develop. Open feedback from our peers and teams helps build on our strengths and develop our less natural skills.

One of the Forties platforms (Forties Alpha) linked this strategy for high involvement to total productive maintenance designed to achieve totally productive operations (autonomous operator maintenance). Other platforms are developing their own systems for successful, results-driven change. The common thread is that they are all based on achieving business results. Diversity is valued, because how do you know you have achieved the best if you only have one experience to measure? Examples of the diverse, results-driven programmes are the adoption of reliability-centred maintenance (RCM) by the Forties Delta team, and Forties Bravo's use of the concepts of zero-based analysis and RCM. We intend to seek out best practice wherever it may exist.

## 12.3 Total productive maintenance

Total productive maintenance is for us, on Forties Alpha, a practical demonstration of empowerment, and enhances our challenge to play a key role in the Forties mission statement.

This is based on three beliefs:

- working together
- winning together
- finishing first.

This is delivered through seven values:

- people
- safety
- quality
- business understanding
- reliable equipment
- empowered teams
- effective teams.

Total productive maintenance (TPM) is the pillar of our future success, since:

1  We all own the plant and equipment.
2  We are therefore all responsible for its availability, reliability, condition and performance within a safe and fit-for-purpose environment.
3  We therefore ensure that our overall equipment effectiveness ranks as the best, by systematically eliminating all non-value-added tasks and reasons for equipment-related waste and loss.
4  We continuously strive for world class performance.
5  We therefore train, develop, motivate and equip our people to achieve these goals by pursuing targeted levels of training hours per period together with a mechanism to evaluate the worth of that training.

The delivery of the TPM message, process and practice is based on:

- the FA TPM strategy and policy
- the sustained commitment and ownership of the TPM process by senior management

- trained TPM facilitators within the platform team
- a carefully tailored TPM awareness and training programme supported by Willmott Consulting Services
- a demanding but realistic three-year to five-year implementation programme based on pilots rolling out to a platform-wide practice of TPM with clearly identified milestones, measurement and exit criteria.

TPM is not a new concept. Indeed, it has been a major success in Japanese industry for approximately 23 years. The dual goal of TPM is zero breakdowns and zero defects, thereby reducing costs and increasing labour productivity. Some Japanese companies that have entered into major TPM programmes have seen 90% reduction in process defects and a general increase in labour productivity of 40%–50%. Since 1987 TPM has been expanding its frontiers and has been enthusiastically adopted in North America, Asia and Europe.

For us, this was a new idea and a new way of working. We were breaking down barriers that had been established over the previous years. First we changed our working practices from single-discipline-based activities to teamwork, focusing on what was best for the asset. Everyone from managers to deck operators had an important role to play. Old taboos were being laid to rest in our drive for excellence.

To take on TPM in our environment was in itself a challenge because nothing like it had been attempted before on an offshore installation. So, you may ask, why is TPM so attractive. Simply:

- It is measurable, visible, practical common sense.
- Employees can understand and therefore value the concept.
- The elements of TPM are not brand new.
- It is a grass roots process.
- It is a world class process.

To set up the process we had to look at the platform and decide on a pilot project that would fulfil our needs. Firstly, everyone had to be involved; secondly, it would have a positive effect on our profitability; and thirdly, it was important that everyone would learn from the pilot so that we could continue to improve in every aspect of our work. With that goal in mind, it was decided to target our produced water disposal (PWD) system. This was a major challenge because of the negative effect it was having on our platform's overall efficiency and our ability to produce maximum oil.

An extensive training programme was undertaken to familiarize the operations teams (onshore and offshore) with the TPM process. Facilitators were trained to assist the teams and coordinate the project. Throughout the project, milestones were set to measure our progress and to ensure that the direction and purpose were maintained.

Each team took ownership of one section of plant with the commitment to deliver its findings within three trips. To succeed in their quest it was imperative that the teams followed the nine-step, three-cycle TPM improvement plan.

The initial benefits delivered from the project were increased teamwork, plant familiarization, and the sharing of experience. However, for the first time many operators and technicians started to have a say in plant optimization and began to have an understanding about equipment effectiveness and operating costs. The team began to discuss availability, performance and quality, and put forward practical suggestions to solve problems. A system and a formula were worked out to measure the equipment effectiveness and to resolve issues that had an undue effect on its performance.

It was important that we followed the improvement plan carefully as experience soon proved that short-cutting the system was counter-productive. Therefore, each team followed the plan and reported some encouraging results.

Everything that we attempted did not come off and we struggled at times to see the benefits. It was costing us both time and money to return the plant to its original condition and to address some of the major issues. We also struggled with understanding our new independence, but generally thrived on the challenge. The biggest percentage of the suggested improvements were simple maintenance activities that were included into our weekly work plan. Other more technical problems were passed to our extended team onshore and their expertise was invaluable in solving some major problems. We were developing into a single focused unit with a desire to succeed. All of our problems as yet have not been solved, but generally speaking we have had a positive response from everyone we have approached. The final challenge is to move from an independent approach to full interdependence across the team, between onshore and offshore and with partners and customers: 'the final stages of adolescence into adulthood.'

It is important with any new system, and especially in the present financial climate, that we get a good return for our investment. Therefore it was important to develop not only an overall equipment effectiveness for the PWD system, but to expand it to cover the overall equipment effectiveness (OEE) of the platform. Our initial results from the PWD system were very disappointing. However now, after a lot of work and development, we are seeing very promising returns on our investment. A field OEE has been worked out and an improvement of 1% on the totals used would cover approximately 30% of our maintenance costs.

## Cost/benefit profiles

If we look at the so-called direct costs of maintenance illustrated in Figure 12.1, there is an inevitable bow-wave effect of implementing a TPM culture (made up of training costs, any additional plant restoration costs and associated time premiums). This bow wave details the fact that TPM drives for zero breakdowns through a greater emphasis on condition monitoring (including operators, who under TPM act as the ears, eyes, nose, mouth and common sense of their maintenance colleagues) and hence maintenance becomes proactive rather than reactive. The bow wave is quite often minimal

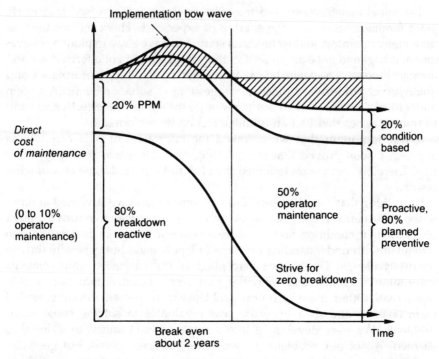

**Figure 12.1**    *BP Forties: cost/benefit profile*

since you spend your existing maintenance budget more effectively from day one on the TPM model and the pilots.

Typical experience of TPM is that it yields significant early individual benefits because of increases in the OEE. These lost opportunity costs of ineffective maintenance, breakdown and plant shutdowns, restart losses, lazy well restimulation, yield losses, water in product, low PWG and loss of safety integrity all have a detrimental effect in terms of the OEE and hence achieving the operating aspirations of Forties Alpha at realistic cost.

One of the main aims of the TPM pilots is to test, measure and improve equipment performance levels and to set stringent but realistic targets for improvement which will show up as bottom-line profit. These gains are made through the TPM problem solving teamwork activities of attacking the lost opportunity costs and then sustaining the improvement by effective asset care and best practice operation and support.

Finally, it was important for us to roll out our plan and develop a continual improvement strategy for the platform. The lessons learnt from each of the five teams need to become the everyday working policy. In addition, the development of the TPM plan should be expanded to cover all of our plant and equipment. As this is a continual improvement plan we have moved a few stages further. Our teams have been remodelled to ensure greater flexibility and cross-training in both problem identification  and solving.

We have a major role to play and TPM has become a vehicle through which our empowered teams can deliver their part to the Forties vision.

No individual group is more or less important than any other. There are no barriers. Information and best practices are shared generously in the knowledge that continuous improvement will give us the high standard of performance that we seek. The roles of the supervisors and managers are no longer directing and controlling. Instead, managers and supervisors provide the self-directed teams with the skills, coaching, technical training and environment required to produce complete success in our business. The work and stature of a supervisor is measured in the number of people he helps to success, not in the number of people he directs and controls.

## 12.4   Plant Manager™: a new system

New initiatives such as empowerment and TPM cannot in themselves deliver all the benefits hoped for. Along with the cultural changes and improved teamwork, new systems and methods of operation that reflect this new environment need to be taken on board. Simply putting a new feel or title to outdated and overly bureaucratic systems will not deliver the performance enhancements now required.

Documentation and information systems were proving to be a barrier to the new approach. It was universally recognized that existing systems fundamentally failed to deliver the quality of information required.

Although rigorous procedures existed to control the documents, the very nature of the system resulted in documentation that always reflected the history of the plant rather than its current status. Previous experience had clearly illustrated that tighter document control did not lead to better integrity of data. The knowledge that needs to be embodied in procedures in order to ensure safety and improve efficiency primarily rests with the personnel who have the day-to-day responsibility for operating the plant.

In line with the new philosophies, an appropriate way to improve and document operating practices is to provide worksite personnel with simple and effective links and tools to collate and manage their own information. This collated and collective knowledge can thus be validated by both peers, leaders and technical authorities.

To this end, Forties engaged the services of Petrotechnics (Aberdeen) with a view to developing an innovative offshore-based system for the offshore collation of relevant procedures, process and instrument drawings (P&IDs) and best practices. This system was to become known as Plant Manager™, designed by Petrotechnics but with Forties operational input.

Given that offshore staff had a major say in the look and feel of the system, the overall response is very much one of: 'Here is a system with input from offshore staff which will be used by offshore staff; we approve.' This is some way distant from the all too common approach of design concept and development being the sole preserve of onshore staff with a little operational input. All too often, the result is: 'Take this system; we know this is what you need.' Experience tells us that this approach has an exceptionally high failure rate. Teamwork was a key element throughout the development of Plant Manager™ and was a major contributor towards the success of the entire project in both content and the delivery method of the end product. In its final incarnation, Plant Manager™ is a Windows™-based application which allows the information associated with systems, subsystems and equipment to be accessed via the simple click of a button.

The program utilizes schematics of the plant as doorways leading to all the procedures and engineering drawings. By clicking on a vessel, the procedures and engineering drawings marked up for a particular process are accessed from a front-end schematic drawing. In addition to delivering and linking information in this manner the system allows the user sufficient privilege to add notes to procedures and annotate drawings without modifying the base engineering drawing itself. Therefore the program provides a cost-effective means of keeping operations documentation safe and up to date. It is a key tool in promoting best practices and safe and efficient operations.

During an offshore trial of an early prototype it became apparent that offshore technicians were very receptive to new ideas and technologies. Isometric sketches of new plant and notes on the latest operating practices were enthusiastically presented for inclusion in the system. The system's ability to allow relevant personnel to add to and amend procedures makes it an ideal source of reference for current best practices – a feature clearly apparent to all those who encountered the prototype model. Developing best practice routines on a continuous improvement basis is one of the cornerstones of TPM.

In addition to the functionality incorporated in the Forties Field implementation, Petrotechnics have continued development, thus allowing the incorporation of all types of media including scanned drawings, photographs, sound, video and links to existing databases (e.g. maintenance management systems). A further innovation is the use of standard CAD drawings as intelligent drawings. Plant Manager™ can utilize these intelligent drawings dynamically with operating procedures and isolation certificates etc. One example of this would be that, as the isolation procedure changes, the drawing will automatically change to reflect the new procedure or command.

## 12.5   Conclusion

Sustaining the improvement has not been easy. 1994 brings new challenges and with it, ever changing demands on our people and equipment.

Key to our continued success is managing our performance as these changes evolve, and that means continuously improving the way we do things.

Empowerment and TPM are no exception. They have given us the environment and tools to make a difference but are not the only solution to all our problems, or the only key to all our opportunities.

Our continued success, therefore, depends on how we manage the change and adapt to meet the new challenges Of all the lessons learnt, this is perhaps the most important and most difficult.

The journey goes on . . . .

## 12.6  Acknowledgements

- To the other platform teams in Forties – it would need another chapter to describe the good things they are doing. We share in their success.
- To all the authors of the many stimulating books and papers on change which challenge the status quo.
- To the people who made the story in this chapter possible. They know who they are. Thank you all.

# ——— 13 ———

# Courtaulds Films Polypropylene

*Tim Holton*, Training and Recruitment Manager

## 13.1 Background to the plant

Courtaulds Films Polypropylene (CFP), based in Swindon and Mantes in France, is part of the Polymers Division of Courtaulds. The Swindon plant employs 311 people and has been manufacturing oriented polypropylene (OPP) for 30 years. It has a turnover of £60 million.

In the 1980s CFP carried out a major capacity expansion at a cost of some £50 million, which added three new lines in Swindon and the acquisition of the French operation to bring the total capacity to around 50 000 tonnes. This took place when growth rates in the industry were of the order of 12–14% a year as OPP was replacing other substrates, particularly cellulose film and paper. The same era saw the development of special films in addition to the commodity coextruded films.

The process involves an extrudate comprising three layers: one relatively thick polypropylene core delivered by tandem extrusion sandwiched between two relatively thin copolymer sealing layers delivered by satellite extruders. The extrudate is combined at a coextrusion die, chilled to get a stable crystalline structure and sequentially stretched in the machine direction, followed by the transverse direction. Film is corona discharge treated and gauged prior to wind-up. It is then slit to customer width, packed on to pallets, warehoused and shipped out. Narrow widths are slit separately in the Customer Slitting Department. A reclaim unit handles all waste film.

Most of Courtaulds' OPP film ends up in the dry food market – crisps and snacks, biscuits, bakery and confectionery. Of our customers, 80% are converters who print, metallize or laminate the film and the rest are end users who use the film without modification.

An MRPII system was installed in 1989 to make sure 95% of the orders reached the customer on time. This was followed by a TQM initiative to tackle the product/quality issues. Although the growth rate has slowed to 6–8% in the 1990s, there is demand in the European markets which has to be fulfilled without extra capacity.

The Swindon plant now consists of four lines with potential capacities of:

line 4    5 000 tonnes
line 5    10 000 tonnes
line 6    9 000 tonnes
line 7    18 000 tonnes

and the French plant has capacities of:

line 1     4 500 tonnes
line 2     5 500 tonnes

There has been a focus on uptime and quality targets because of a history of instability in manufacturing, and hence now a focus on TPM in order to unlock the potential capacity.

## 13.2   Operations restructure to become competitive

CFP reorganized its operations structure at Swindon on 1 January 1992 to make it leaner, more flexible and better skilled. It then became a base to instigate change more quickly.

The main features of the reorganization were:

1   The plant was divided into six separate units: line 4, line 5, line 6, line 7, Reclaim and Services, and Customer Slitting (Figure 13.2).
2   Each unit had a dedicated team on shift to achieve ownership and accountability.
3   There was a separate maintenance team on shift servicing all units.
4   All teams were led by working team leaders who reported to a shift manager.
5   Each unit had a dedicated unit team on days comprising specialist engineers and technologists who would be responsible for performance, customer service and engineering support.
6   There would be flexibility of skills within the shift teams to get away from the one-man one-job principle. Each skill had an associated payment.
7   The QC department was disbanded and shift teams did their own QC.
8   Nineteen job titles were reduced to four, i.e. team leader, process technician, craft technician and material controller.
9   Process technicians were able to gain simple mechanical and electrical skills, and craft technicians could gain operations skills as well as cross-function skills, i.e. mechanical to electrical and electrical to instrument technician.

| | Line 5 | Line 4 | Line 7 | Line 6 |
|---|---|---|---|---|
| Width (m) | 6 | 4.5 | 8 | 6 |
| Capacity (t/a) | 10 000 | 5000 | 18 000 | 9000 |
| Product | Pearlized and white specials | Pearlized and white specials | Commodity films | Thin clear specials |
| Reclaim | five extrusion lines | | | |
| Customer slitting | five machines | | | |

**Figure 13.2**   *Courtaulds: Swindon plant summary*

10  The shift system was changed from five crew to six crew, with each shift working a continental-style eighteen-week cycle, rotating through shifts and days. The change was necessary to carry out essential flexibility training on days without incurring overtime.
11  The appraisal system was extended to process and craft technicians and called a personal development review.
12  Job numbers were reduced.
13  New terms and conditions of employment were negotiated jointly with both on-site unions.

A fine tuning of this structure was made in March 1993 to focus the factory into two reporting structures, i.e. lines 4 and 5, separate from lines 6 and 7 and Reclaim. Instead of a shift manager controlling the whole factory, two business team leaders were created, one each side of the factory, reporting to two separate business managers. Business team leader is still a working supervisory position.

## 13.3    Formation of Teams

Although we attempted to utilize teamwork for the TQM initiative in 1991, it was generally spasmodic and only involved a small fraction of individuals on the shop floor. The problem was mainly trying to form multi-disciplinary teams with people from different shifts, using a complicated route-map to solve product quality-related problems.

Teamwork training started in earnest in April 1992 when we trained the team leaders. Each shift had six team leaders and they were given leadership training via task-related activities at the Courtaulds Training Centre, Kenilworth. The added value of this training was twofold. Firstly, it cemented a team within a shift team which carried out improvement projects on their shift; and secondly, the team leaders requested similar training for their individual teams. The outcome was a second series of training events called team building, and this took place mainly outdoors on Salisbury Plain. The individual teams were now ready to carry out some form of project work on their shift. TPM was therefore an ideal initiative for the teams to tackle.

## 13.4    Need for TPM

The need for TPM in a capacity-constrained business is obvious. We have the flexibility of our workforce to ensure lines do not stop, but no control over the breakdowns or the deficiencies in the equipment which cause quality problems. Speeds have been lowered to ensure uptimes increase to the target levels. We estimate there are 300 breakdowns a month which cause lines to stop. This does not include major stoppages to slitting or reclaim equipment.

One of the lines was pulled up in 1992 for an extensive two-week refurbishment (top to bottom) in order to increase its uptime. Had TPM been in place, this exercise would not have been necessary.

We have high-tech equipment which needs regular asset care by the people who work on the machines so that costly refurbishments or replacements are not necessary. As a supplier to the food industry, we also need a clean and tidy plant to work in. This requires a dust-free environment.

We also want to be number 1 in coextruded OPP. If this means being a so-called world class manufacturer, then so be it. TPM we believe will give long-term gains in all these areas, as well as a feeling of ownership and pride in the workforce.

As a first stage, we have focused on uptime (90% for all lines) but this will be superseded by OEE with a three-year target of 85%.

## 13.5  What is already in place to help the TPM initiative

*Teams, communication*

In addition to teams and trained team leaders who are willing to take initiatives, we have a flexible workforce, some of which we have already trained in mechanical and electrical skills. Team building is linked into the business objective which requires each person to achieve his/her full potential in a safety conscious environment. We believe TPM will enhance teamwork still further.

The Chief Executive communicates the business objective every six months directly to all the workforce, and team briefing is cascaded down monthly to all the workforce. This has removed suspicion from employees. Good communication and a cooperative union are prerequisites to make an initiative like TPM happen. One shift pattern for the majority of the workforce, plus the sixth shift, are also an added bonus. An initiative tried in 1986 failed because poor communication led to union non-cooperation.

*Maintenance systems*

R&Ms are carried out on each line and slitter every ten to twelve weeks. The routines include checks, measures, routine replacements, greasing, lubrication, urgent jobs and minor modifications. The R&M is planned by the unit teams and carried out by craft technicians, unit engineers and contract labour.

There is a computer maintenance system which records and analyses all maintenance activities. Some predictive maintenance is carried out on line 6 by condition monitoring (vibrational analysis).

## 13.6   Potential barriers and TPM success

*Craft technicians*   TPM is seen as a threat to their jobs in a lean company.
*Unit teams*   The barrier here is 'we know best' when it comes to process technicians taking initiatives on their equipment.
*Process technicians*   'Flavour of the month': another three-letter acronym.
*Directors*   A better initiative is just around the corner.
*Senior managers*   Another initiative taking my valuable time resource.
*Craft and process*   Separate empires.
*Craft and process*   A novel way to improve cleaning.

## 13.7   How we introduced TPM

After contracting Willmott Consulting Services, we followed the Western approach to TPM to pilot the TPM process. The key steps were:

1   Awareness for senior managers: February 1993.
2   Scoping study (attitude survey of a cross-section in operations): March.
3   Commitment from senior managers to proceed: March.
4   Awareness session for key personnel.
5   Select steering group, project champion, site facilitator and facilitators.
6   Select pilot project areas (Figures 13.3, 13.4).
7   Train eight facilitators: April.
8   Train squad support: April.
9   Awareness for shifts, select teams and start eight pilot projects on the twelve-week cycle: April–May.
10   Monitor progress through steering group meetings: April–July.
11   Clear and clean areas started: June–July.
12   Japan study tour for four personnel: June.
13   Presentation of eight pilot projects: June–August.
14   Review of pilot project: September.
15   Plans for two-stage roll-out: start October.

## 13.8   Potential benefits of TPM

Some historical OEE values are shown in Figure 13.5. The model for April 1993 shows the results in Figure 13.6. Each 1% is worth £105 000 per year on availability alone.

## 13.9   Conclusions

TPM is a long journey with a start point but not a finish: you just keep getting better. It therefore has to become a habit, just like working safely, which carries on regardless of changes to an organization. We at CFP have just started on the journey.

TPM project suitability selection

1. Draw schematic of overall project area, identifying significant functional blocks.
2. Write the block names in 'Project area' column.
3. Assess each 'Project area' as suitable under the five headings.
   Score in range 1 to 3: 1 = low, 3 = high.
4. Suggest you select project area with highest overall score unless there are other factors that lead you to another decision. If so these must be strong and clear reasons that are accepted by team and facilitator.

| Project area | What will be the impact on OEE? | Can we measure the change in OEE? | Visual impact in the area? | Applicable to the full TPM implemen-tation (1 to 9)? | Will it encourage teamwork between production and maintenance | Total |
|---|---|---|---|---|---|---|
| | | | | | | |
| | | | | | | |
| | | | | | | |
| | | | | | | |
| | | | | | | |
| | | | | | | |
| | | | | | | |
| | | | | | | |
| | | | | | | |
| | | | | | | |
| | | | | | | |
| | | | | | | |
| | | | | | | |
| | | | | | | |
| | | | | | | |
| | | | | | | |
| | | | | | | |
| | | | | | | |
| | | | | | | |
| | | | | | | |
| | | | | | | |
| | | | | | | |
| | | | | | | |
| | | | | | | |
| | | | | | | |
| | | | | | | |
| | | | | | | |
| | | | | | | |

**Figure 13.3** *Courtaulds: TPM project selection form*

| Line 6 | Extrusion | Delta |
|---|---|---|
| Line 6 | Prime slitter | Alpha |
| Line 4 | Extrusion | Charlie |
| Line 5 | Treater | Foxtrot |
| Line 7 | Prime slitter | Echo |
| Reclaim | | Bravo |
| Customer slitting | | |

**Figure 13.4**   *Courtaulds: selected pilot project areas*

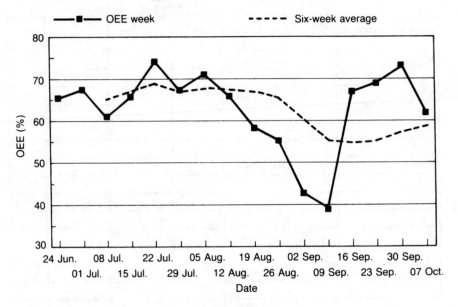

**Figure 13.5**   *Courtaulds: line 6 historical OEE*

| | OEE (%) | | Availability (%) | | Performance rate (%) | | Quality rate (%) |
|---|---|---|---|---|---|---|---|
| Current | 54 | = | 85 | × | 75 | × | 85 |
| Best of best | 81 | = | 95 | × | 90 | × | 95 |
| World class | 86 | = | 95 | × | 95 | × | 95 |

**Figure 13.6**   *Courtaulds: line 6 OEE April 1993*

# TPM Success Stories from the Automotive Industry

## 14.1 Case study: Automotive manufacturer 1

*Business issue*

How to achieve and sustain world class manufacturing status through JIT and lean production.

*Key success factors*

Recognition that JIT and lean production philosophies only work if quality output can be delivered from highly reliable and effective manufacturing processes and equipment. Also that consistent and sustainable levels of equipment performance are achieved by operators and maintainers rather than systems solutions.

*Impact of TPM*

On two TPM pilot equipments, the overall equipment effectiveness performance indicators in Figure 14.1 have been set. The respective TPM pilot teams have an identified improvement programme to achieve the targets within six months.

|  | Cell A | Cell B |
|---|---|---|
| Current OEE | 81% | 63% |
| Best of best | 95% | 88% |
| Increase output | +18% | +55% |
| *or* | | |
| Fewer hours | −16% | −28% |

**Figure 14.1** *Case study: pilot equipment TPM potential*

*Quotable quotes*

> I was supicious about the motives of TPM originally, but I can now see it's
> benefits driven. So for us at least, an increase in our OEE equals job security.
> TPM team member

## 14.2   Case study: Automotive manufacturer 2

*Business issues*

How to achieve customer satisfaction, cost competitiveness and world class
performance.

The production system is the key driver to achieve this vision through:

- reduced cycle time
- continuous improvement
- one-piece flow
- total productive maintenance.

*Key success factor*

The increasing sales volume of off-road and small commercial vehicles has
stretched production capability. The door robot is a bottleneck welding
centre using highly sophisticated jigs to secure a range of door profiles.
The single-arm Fanuc robot is programmed to suit any product mix but
the throughput is dependent on the combination of the jig and robot working effectively.

*Impact of TPM*

Achievement of best of the best (Figure 14.2) is worth an additional £100 000
sales value per year, through TPM-driven shop floor improvements and
sustained best practice.

Achievement of 85% levels of OEE will be worth an additional £215 000
sales value per year.

|  |  | Added value per year |
| --- | --- | --- |
| Current OEE | 53% |  |
| Best of best | 68% | £100 000 |
| Future requirement | 85% | £215 000 |

**Figure 14.2**   *Case study: welding centre TPM potential*

*Quotable quotes*

> The value of TPM in my area is unquestionable. Getting our equipment right through TPM will avoid diluting the benefits of our other initiatives.
>
> Unit Manager

> That'll do . . . It's good enough . . . It's working so leave it alone. These sorts of attitude must never be accepted if we are to become a world class organization. If used effectively, TPM could be the most significant change to affect production and maintenance since Japan's entry into the car market.
>
> Manager Continuous Improvement

## 14.3   Case study: Automotive manufacturer 3

*Impact of TPM*

For a gear cluster line/cell, consistent achievement of best of the best (80% versus 56%) was worth 360 extra pieces per shift, or the same output per shift in $2\frac{1}{2}$ hours less. This was achieved by *resolving* identified problems.

Meanwhile minor stoppage response time was reduced from 2.75 minutes per event to 1.20 minutes to give 50% of this potential, immediately and at no cost.

*Quotable quotes*

> TPM is making rapid inroads into our reliability problems because of the structured approach which our consultants (WCS) have introduced. In the past our experience has been that consultants have shown us concepts and we have had to work out how to apply them. This is a much more practical and hands-on approach.
>
> Head of Continuous Improvement Initiative

> The overall equipment effectiveness ratio is the most practical measure I have seen.
>
> Senior Manager

> TPM is an excellent team building process which helps develop the full potential of our people.
>
> Head of Maintenance

> TPM plugs nicely into our TQM programme.
>
> TQM Facilitator

> Change initiated by the team (through TPM) is more rapidly accepted into the workplace than when imposed by management.
>
> TPM Champion

# Index